AC AND DC
MOTOR CONTROL

AC AND DC
MOTOR CONTROL

Gerald A. Moberg

Teaching Master
Algonquin College
  of Applied Arts and Technology
Nepean, Ontario, Canada

John Wiley & Sons
New York, Chichester, Brisbane, Toronto, Singapore

Photographs and illustrations in the text that are identified by this logo appear courtesy of Allen-Bradley Canada Limited.

 Photographs and illustrations in the text that are identified by this logo appear courtesy of Square D Canada.

Library of Congress Cataloging-in-Publication Data:

Moberg, Gerald A.
AC and DC motor control.

Includes index.
1. Electric controllers. 2. Electric motors—
Starting devices. I. Title.

TK2851.M63   1987   621.46′2   86-11074
ISBN 0-471-83700-8

Printed in the United States of America

10 9 8 7 6 5 4 3 2 1

To my dear wife Vera Evelyn,
whose inspiration and encouragement
made this textbook possible

PREFACE

The amount of information a tradesperson in the electrical industry is exposed to, can be monumental. With this in mind, AC and DC motor control, relay logic, and the related electrical code are presented in a manner easy to understand, and applicable to on-the-job situations.

The theory of control, operation, and design is presented clearly and concisely, and a logical approach is taken to instruct the reader ''how-to'' and ''why.''

The threads that tie the various sections of the textbook together are the hands-on approach and the requirements of the electrical code. Although the reader needs a working knowledge of basic electrical theory and terminology before making use of this textbook, the mathematical requirements are minimal.

Section 1 provides an introduction to motor control, and Section 2 sets the stage for the application of the electrical code requirements. Section 3 provides a detailed study of full voltage motor starting. Section 4 entitled ''Specialty Circuits'' discusses circuits and equipment that are of utmost importance. The material discussed in Section 5 is common to Sections 6 to 11, introducing time-delay and transition.

Sections 6 to 10 deal with the design, installation, and troubleshooting of power and control circuits for the various starting methods used to start large AC motors. Section 11 covers design, installation, and troubleshooting of power and control circuits for DC motors.

Summary questions have been included for all sections, with electrical code, troubleshooting, and multiple-choice problems provided for Sections 6 to 11.

Lab Sections A and B comprise the Student's Lab Manual and have been carefully designed to coincide with the lessons presented. The manual provides the opportunity to apply the concepts of the text discussion in a meaningful way.

The solutions to the lab assignments are included in an Instructor's Manual. It is the instructor's prerogative to accept, add to, or delete any particular assignment shown in the student's manual. The text and lab manual may be used at various levels of instruction.

A discussion of solid state control and programmable controllers is purposefully omitted as these topics require in-depth coverage and should not be considered lightly. The subject of relay logic must be fully understood before moving on to programmable control.

In addition to being used in the classroom, this textbook will be useful as a reference by electricians, refrigeration technicians, maintenance personnel, electrical inspectors, electrical designers, and anyone involved with the design, installation, or maintenance of motor control and related equipment.

GERALD A. MOBERG

ACKNOWLEDGMENTS

I express my appreciation to the individuals and organizations who kindly permitted the many electrical circuits and photographs to be used in this textbook. In particular, I thank W. C. Torrance, Vice President—Marketing, Allen-Bradley Canada Limited, Cambridge, Ontario, and D. B. Langford, Commercial Vice President–Canadian Marketing, Square D Canada, Mississauga, Ontario, for the kind cooperation extended to me. Thanks are also extended to F. Kuntz, Director of Operation, Standards Division, Canadian Standards Association, Toronto, Ontario, for permitting a portion of the Canadian Electrical Code to be included in the text.

Special thanks must go to John M. Paul, CET, Senior Electrical Designer and Director, J. L. Richards and Associates Limited, Consulting Engineers and Planners, Ottawa, Ontario for encouragement during the writing of the text.

I also express my gratitude to the editor, Hank Stewart and to the many hard working members of the staff at John Wiley & Sons who helped to make this book a reality: Elizabeth Doble, Frank Grazioli, Deborah Herbert, Lilly Kaufman, Joe Keenan, Ishaya Monokoff, Kevin Murphy, and Kieran Murphy.

Without the encouragement and cooperation from my family, this book could never have been written. For this I thank my wife, Vera Evelyn, and my two sons, Ray and Don, who are both electricians.

G. A. M.

CONTENTS

STUDENT LAB MANUAL

AC AND DC
MOTOR CONTROL

SECTION 1

MOTOR CONTROL

INTRODUCTION TO MOTOR CONTROL

Motor control plays an important part in industry today. Industries would cease to function without properly designed, coordinated, installed, and maintained control systems. Motor control is that portion of the electrical system which starts, stops, and reverses motors that drive various pieces of equipment. In addition, motor control equipment is designed to limit starting current and control starting torque.

The control systems must be designed and installed to provide the correct sequence of operation for the driven equipment. The electrical engineer designing the control system will follow the requirements of the electrical code. In addition to proper equipment operation, care must be taken to ensure the safety of the maintenance personnel by supplying adequate disconnecting means.

Electricians, maintenance personnel, and service technicians are required to glean control system information from the engineering drawings and apply it to the actual work situation at hand.

This section is intended to present an overview of the subjects of magnetic motor control and the electrical code. Each topic will be explored in detail in succeeding sections.

SYMBOLS

Motor control language consists of symbols to express an idea or to form a circuit diagram, which may be understood by trained personnel. The symbols illustrated in Figures 1-1 to 1-12 are those most commonly used in industry today. Each symbol will be discussed in detail throughout the text.

o—∿o Single-pole switch

Three-pole switch

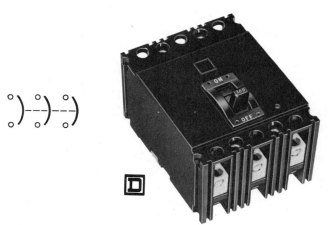

FIGURE 1-1
Three-Pole Fused Switch.

FIGURE 1-2
Three-Pole Circuit Breaker.

⌇ Fuse

FIGURE   1-3
Overload Relays, Heaters, or Elements.

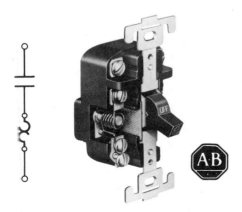

FIGURE   1-4
Single-Pole Manual Motor Starter.

FIGURE 1-5

Three-Phase Magnetic Motor Starter.

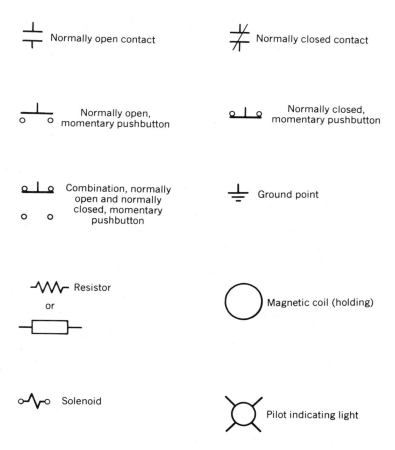

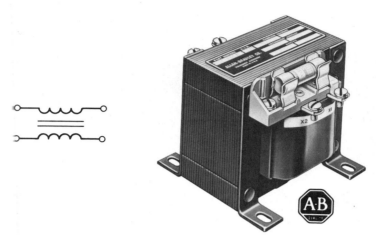

FIGURE 1-6
Control Transformer.

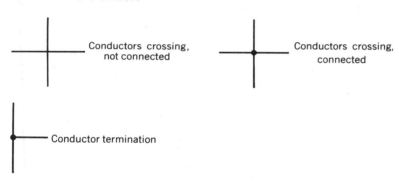

Conductors crossing, not connected

Conductors crossing, connected

Conductor termination

(a) Normally closed

(b) Normally open

(c) Held open

(d) Held closed

FIGURE 1-7
Limit Switch.

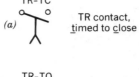

TR-TC

(a) TR contact, timed to close

TR-TO

(b) TR contact, timed to open

FIGURE   1-8
Pneumatic Timer (with coil).

TS-*

(a) Triggered switch, * coil identified

TS-*

(b) Triggered switch, * coil identified

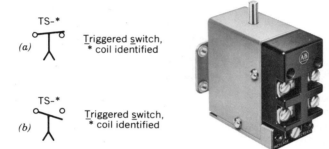

FIGURE   1-9
Pneumatic Timer (triggered contacts).

(a) Normally closed (identify, will open or close on temperature rise)

(b) Normally open (identify, will open or close on temperature rise)

FIGURE   1-10
Temperature Actuated Switch.

(a) Normally closed

(b) Normally open

FIGURE 1-11
Pressure- and Vacuum-Actuated Switch.

Foot Switch

(a) Normally open

(b) Normally closed

Liquid Level Switch

(a) Normally closed

(b) Normally open

Flow Switch

(a) Normally closed

(b) Normally open

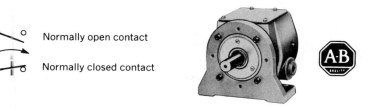

Normally open contact

Normally closed contact

FIGURE 1-12
Zero-Speed Switch.

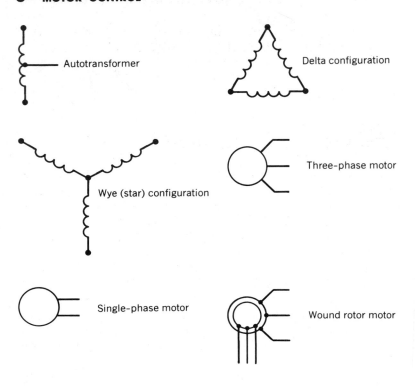

As a further introduction to motor control, a few important terms must be mentioned at this point. Terminology is of the utmost importance when attempting to understand the electrical code and motor control.

OVERCURRENT PROTECTION VERSUS OVERLOAD PROTECTION

In the study of motor control, two terms must be understood and used correctly. The terms overcurrent protection and overload protection will form the thread that ties together the study of the electrical code throughout this text. Overcurrent protection is installed in the power circuit to protect the conductors and may exist in the form of fuses or circuit breakers, while overload protection is installed in the circuit to protect the motor windings and may exist in the form of overload relays, heaters, or elements. Each of these terms will be discussed in detail throughout the text as required.

WIRING DIAGRAM VERSUS SCHEMATIC DIAGRAM

A *wiring* diagram attempts to illustrate the physical location of all components. Coils, contacts, motors, and the like are shown in the actual position that would be found on an installation. A wiring diagram can make it easier to determine the required number of conductors between points in a circuit, but it becomes difficult to trace the circuit.

A *schematic* diagram should be used when designing or troubleshooting an installation. Control components are rearranged to simplify the tracing of the circuit. Line, ladder, or elementary diagrams are other terms used in lieu of a schematic. A tradesperson must develop the ability to translate a wiring diagram into a schematic, and a schematic into a wiring diagram.

This text will assist one in the art of designing and reading schematic diagrams.

POWER CIRCUIT VERSUS CONTROL CIRCUIT

Power circuits and control circuits are the topics that form the backbone of this text. Power circuit conductors are sized according to the current drawn by the motor and form the power circuit, while the control circuit contains the control devices that will initiate the starting or stopping of the motor. Again, each of the circuits will be studied in detail as the topics arise.

The single-line drawing shown in Figure 1-13 is a method used by many motor control equipment manufacturers as a road map in the study of motor control installations. The single-line drawing will aid the reader in understanding the terminology used throughout the electrical code, motor control manuals, and this text.

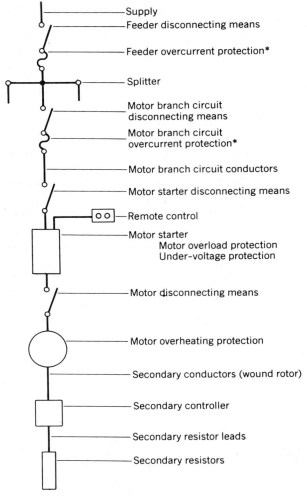

Supply
Feeder disconnecting means

Feeder overcurrent protection*

Splitter

Motor branch circuit
disconnecting means

Motor branch circuit
overcurrent protection*

Motor branch circuit conductors

Motor starter disconnecting means

Remote control

Motor starter
Motor overload protection
Under-voltage protection

Motor disconnecting means

Motor overheating protection

Secondary conductors (wound rotor)

Secondary controller

Secondary resistor leads

Secondary resistors

*Fuses or circuit-breakers

FIGURE   1-13

**Single-Line Drawing of Typical
Motor Installations.**

MOTOR STARTER AND CONTROL EQUIPMENT SELECTION

Care must be exercised when selecting motor control equipment. The procedure for selecting the correct equipment must start with the motor starter size. The following chart (Table

TABLE 1-1

Horsepower _____
Voltage _____
Phase _____ _____ Size ___
Hertz _____

1-1) lists the information required to determine the motor starter size.

Motor Starter Ratings

Motor starters are rated in horsepower and are assigned a NEMA (National Electrical Manufacturers Association) or an EEMAC (Electrical Electronics Manufacturers Association of Canada) number. The size numbers for voltages 200 to 600 V range from 00 to 9. Refer to the chart shown in Table 1-2. The size number may be selected by applying the horsepower, voltage, and phase to the chart.

Derating Motor Starters

Jogging is defined as an operation in which the motor runs when the pushbutton is pressed and will stop when the pushbutton is released. Jogging is used when the motor must run for short periods of time to allow a machine setup. Another term for jogging is inching.

Plugging is a method of stopping a polyphase motor quickly by momentarily connecting the motor for the reverse rotation while the motor is running.

Jogging or plugging causes excessive heating of the power circuit contacts; therefore, if a motor is jogged more than five times per minute or if plugging is used, the rating of the motor starter must be derated.

Compare the NEMA or EEMAC ratings for the motor starter used for jogging, as indicated in Table 1-3, with the ratings shown for full voltage starting in Table 1-2.

TABLE 1-2

Motor Starter Rating Chart for Full Voltage Starting

EEMAC or NEMA	Maximum hp Ratings		
	Motor Volts	Single-Phase	Three-Phase
00	200	1	$1\frac{1}{2}$
	230	1	$1\frac{1}{2}$
	460	—	2
	575	—	2
0	200	2	3
	230	2	3
	460	—	5
	575	—	5
1	200	3	$7\frac{1}{2}$
	230	3	$7\frac{1}{2}$
	460	—	10
	575	—	10
2	200	$7\frac{1}{2}$	10
	230	$7\frac{1}{2}$	15
	460	—	25
	575	—	25
3	200	15	25
	230	15	30
	460	—	50
	575	—	50
4	200	—	40
	230	—	50
	460	—	100
	575	—	100
5	200	—	75
	230	—	100
	460	—	200
	575	—	200
6	200	—	150
	230	—	200
	460	—	400
	575	—	400
7	230	—	300
	460	—	600
	575	—	600
8	230	—	450
	460	—	900
	575	—	900
9	230	—	800
	460	—	1600
	575	—	1600

TABLE 1-3

Motor Starter Rating Chart for a Motor Used for Jogging or Plugging

Phase	Motor Volts	Plugging Horsepower Rating						
		Size 0	Size 1	Size 2	Size 3	Size 4	Size 5	Size 6
3	200	$1\frac{1}{2}$	3	$7\frac{1}{2}$	15	25	60	125
	230	$1\frac{1}{2}$	3	10	20	30	75	150
	460	2	5	15	30	60	150	300
	575	2	5	15	30	60	150	300

Motor starter sizes for

A. Primary resistance starting.
B. Wye-delta starting.
C. Part-winding starting.
D. Autotransformer starting.
E. Wound rotor starting.

will vary from the ratings for full voltage starting, as shown in Table 1-2. Check the manufacturer's motor manual to determine proper sizes.

Having completed the first step, which was to determine the starter size, move on to the final steps.

The following is a list of factors to consider when selecting motor control equipment.

A. Reversing.
B. Nonreversing.
C. No voltage protection circuit.
D. No voltage release circuit.
E. Across-the-line starting.
F. Reduced voltage (current) starting.
G. Multispeed.
H. Multispeed across-the-line.
I. Multispeed reduced voltage (current) starting.
J. Manual starter.
K. Magnetic starter.

It can be readily seen that many factors enter into the selection of control equipment.

Each component selected must have an enclosure approved for the environment to which it will be exposed.

Each component must also be approved for

A. The voltage to which it will be connected.

B. The current that it must carry.

C. The horsepower that it must control.

SAFETY

Safety on the job is a requirement often thought to be the other person's problem. Each person, however, is ultimately responsible for his or her own safety and well-being. Some will refer to safety rules in a lighthearted way, saying "rules are made to be broken," but safety rules have been instituted and enforced for the purpose of protecting individuals from accidental injury. Always keep in mind that carelessness can result in harmful accidents.

Studying electric motor control from a textbook will prove to be a nonhazardous occupation. When the ideas obtained from a text are put into practice, the subject is better understood, but the student or tradesperson may still be exposed to hazardous conditions. On-the-job experience will expose the worker to "live" equipment and rotating machinery.

Before proceeding to work on a piece of equipment, disconnect that equipment from its electrical supply system. Check the schematic to ascertain whether the machine is interlocked with another piece of equipment. If it is, two or more electrical circuits may be interconnected. Lock out and tag each disconnecting means, using locks that can be unlocked only by the installer. Using a voltage tester or meter, check for potential before carrying out the necessary repairs. Do not assume the circuit is safe to approach until all testing has been completed. Many maintenance personnel have been injured while working on a so-called "dead" circuit.

Do not:

• Depend on pilot lights to indicate when the supply voltage has been disconnected.

- Provide extra keys for your personal lockout device.
- Assume the circuit is dead and safe to work on.

Do:

- Follow lockout procedures as recommended by the accident prevention authority having jurisdiction in your particular area.

Care and common sense can make the working area safe. Do not take chances. **BE ALERT. THINK SAFETY.**

METRIC SI

SI is the abbreviation in all languages for the Système International d'Unites (International System of Units). The electrical industry will eventually convert entirely to this system of measurement. Units affected in this text by the change are volts, amperes, and inches. The units for volts and amperes remain unchanged. Conduit diameters previously measured in inches will change to millimeters (mm). To convert from inches to mm, simply divide inches by 0.03937. When applicable, both dimensions have been shown here.

ELECTRICAL CODE INTERPRETATION

Throughout this text, reference will be made to the *electrical code*. The electrical code pertinent to you is the one that governs electrical installations in your particular area.

Electrical codes are subject to interpretation, but much research has been done to arrive at one accurate, acceptable interpretation. To better understand this text and motor control in general, your electrical code should be used in conjunction with the textbook.

Topics that involve the electrical code are

1. Disconnecting means
      rating.
      installation.
2. Motor starters
      rating.
      installation.

3. Control equipment.
4. Overcurrent protection.
5. Overload protection.
6. Conductors.
7. Raceways.
8. Safe operation of electrical equipment.

The codes used in your area should be compatible with the interpretations given throughout this text. If a difference in interpretation is noted, keep in mind that the interpretation of your particular inspection authority or agency must be adhered to.

REVIEW EXERCISES

1-1 Define "overcurrent protection."

1-2 Define "overload protection."

1-3 Define "power circuit."

1-4 How is a motor starter rated?

1-5 What designation is given to a motor starter to indicate rating?

1-6 What do the letters "NEMA" represent?

1-7 What does "EEMAC" stand for?

1-8 Why must motor starters be derated if the circuit is used for jogging or plugging?

1-9 Discuss safety procedures pertaining to electrical equipment.

1-10 Discuss the purpose for, and the location of, a disconnecting means in a motor installation.

1-11 Spend time with the electrical code book, studying topics such as motor control and wiring methods.

SECTION 2

APPLIED ELECTRICAL CODE

INTRODUCTION TO THE ELECTRICAL CODE

This section will consist of the study of the electrical code requirements for the power circuits for full voltage starting of squirrel-cage motors. The study of full voltage starting will be covered in Section 3, but for the purpose of the electrical code requirements, let us define full voltage starting as an installation in which a motor is started by connecting the motor directly across full line voltage.

The topics discussed in this section must be altered to suit the various starting methods shown in Sections 6 to 11. The procedure will remain the same, but the percentages for the overcurrent devices will vary, and the method of selecting the overload devices must be understood.

In order to continue with the study of the electrical code as it pertains to motor control, the following terms must be introduced:

A. Disconnecting means.
B. Overcurrent protection.
C. Overload protection.

Disconnecting Means

The disconnecting means is the device which may be operated that will isolate the electrical equipment. The disconnecting means must be installed so that there will be no hazard to the tradesperson who works on the installation.

The electrical code divides the subject as follows:

A. Disconnecting means required.
B. Types of disconnecting means.
C. Rating of disconnecting means.
D. Location of disconnecting means.
E. Accessibility of disconnecting means.

Each of the above must be considered when designing an installation. Think safety.

Overcurrent Protection

Overcurrent protection is installed in the circuit to protect the conductors and may exist in the form of fuses or circuit breakers.

Overcurrent protection devices are sized and selected based on the full-load amperes (FLA) of the motor. Care must be exercised when selecting the overcurrent device. The maximum overcurrent device will be a percentage of the FLA of the motor. A careless selection of a fuse or circuit breaker may result in damage to the wiring system and possibly injury to personnel. The electrical code is very specific in regard to the selection of overcurrent devices.

Overload Protection

Overload protection is installed in the circuit to protect the motor and may exist in the form of overload relays, heaters, or elements. The electrical code divides the subject as follows:

A. Overload protection required.
B. Overheating protection required.
C. Types of overload and overheating protection required.
D. Number and location of overload protective devices.
E. Shunting of overload protection during starting.

TABLE 2-1

Motor Overload Device Requirements

System	For Motor Overload Protection		Kind of Motor
		Number and Location of Overload Devices	
Three-wire, three-phase, AC	Three, one in each phase		AC, three-phase
Four-wire, three-phase, AC			
Single-phase, two-wire	One in any conductor except a neutral or grounded conductor		Single-phase

F. Automatically started motors.

G. Automatic restarting after overload.

The overload protection is designed to protect the motor windings when the motor is running, and not during the start-up period. At this point, check the topics listed in the electrical code book.

Table 2-1 shows, at a glance, the number of overload devices required for various power supply systems.

The term "neutral" means that the conductor (when one exists) of a polyphase circuit, or of a single-phase three-wire circuit, is intended to have voltage such that, the voltage differences between it and each of the other conductors are approximately equal in magnitude and equally spaced in phase.

However, in the trade, the term "neutral conductor" is commonly applied to that conductor of a two-wire circuit that is connected to a conductor, which is grounded at the supply end.

The approach taken in this text is the practical "trade" approach; we refer to the grounded circuit conductor as a "neutral" conductor. This does not violate any codes and should be easier to relate to the circuits used throughout.

MOTOR SERVICE FACTOR

If a motor manufacturer has given a motor a "service factor," it means that the motor can be allowed to develop more than its rated or nameplate horsepower, without causing undue deterioration of the insulation. The service factor represents a margin of safety. If, for example, a 25 hp motor has a service factor of 1.15, the motor can be allowed to develop 28.75 hp. The service factor depends on the motor design.

If the service factor of a motor is 1.15, the current rating of the overload device must not exceed 125% of the nameplate current rating of the motor. If the service factor of a motor is 1.0, the current rating of the overload device must not exceed 115% of the nameplate current rating of the motor.

The motor with a 1.15 service factor could relate to a heavy-duty motor, and as such, the electrical code will permit the motor control manufacturer to design an overload device rated at 125% of the motor nameplate current. Conversely, a motor with a 1.0 service factor may have an overload device rated at not more than 115%.

BRANCH CIRCUIT OVERCURRENT PROTECTION

The three most common overcurrent devices are

A. Time-limit type circuit breaker.
B. Time-delay "D" fuse.
C. Non-time-delay fuse.

To select the maximum size overcurrent device for a given motor branch circuit, apply the nameplate current rating of the motor to the respective table in the electrical code book. The maximum rating of the overcurrent device will be shown opposite the FLA and under the appropriate column.

The table is based on percentages listed on a second table in the electrical code book. Table 2-2 shows such a table for full voltage starting.

CONDUCTORS

The electrical code recognizes and accepts the installation of both copper and aluminum conductors. If properly installed and terminated, either installation will be accepted. Throughout this text, copper conductors have been used in order to standardize the selection of conductors and conduits.

TABLE 2-2

Canadian Electrical Code, Part 1

(TABLE 29)

(See Rules 28-200, 28-204, 28-208, and 28-210*)

RATING OR SETTING OF OVERCURRENT DEVICES FOR THE PROTECTION OF MOTOR BRANCH CIRCUITS

Type of Motor	Percent of Full-Load Current		
	Maximum Fuse Rating		Maximum Setting Time-Limit Type Circuit Breaker
	Time-Delay** "D" Fuses	Non-Time Delay	
Alternating Current			
Single-Phase, all types	175	300	250
Squirrel-Cage and Synchronous:			
Full-Voltage, Resistor and Reactor Starting	175	300	250
Auto-Transformer Starting:			
Not more than 30 A	175	250	200
More than 30 A	175	200	200
Wound Rotor	150	150	150
Direct Current	150	150	150

* (Except as permitted in Table 26 where 15-A overcurrent protection for motor branch-circuit conductors exceeds the values specified in the above table.)

** *Time delay "D" fuses are those referred to in Rule 14-200.*[a]

NOTES: (1) *The ratings of fuses for the protection of motor branch circuits as given in Table 26 are based on fuse ratings appearing in the table above, which also specifies the maximum settings of circuit breakers for the protection of motor branch circuits.*

(2) *Synchronous motors of the low-torque low-speed type (usually 450 rpm, or lower) such as are used to drive reciprocating compressors, pumps, etc., and which start up unloaded, do not require a fuse rating or circuit-breaker setting in excess of 200% of full-load current.*

(3) *For the use of instantaneous trip (magnetic only) circuit interrupters in motor branch circuits see Rule 28-210.*[a]

[a] The (Table 29) and the Rule numbers mentioned refer to the Canadian Electrical Code. Similar comments will be listed in the National Electrical Code. With permission of Canadian Standards Association, this table is reproduced from CSA Standard C22.1-1982, Canadian Electrical Code, Part 1 (14th edition), which is copyrighted by CSA; copies may be purchased from CSA, 178 Rexdale Blvd., Rexdale, Ontario M9W 1R3.

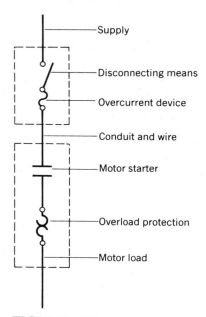

FIGURE   2-1

Single-Line Drawing for the Single Motor Installation Example.

POWER CIRCUIT CALCULATION

Figure 2-1 shows a single-line drawing of the power circuit for a single motor installation. The example shows the procedure to be followed when designing the power circuit. Figure 2-2 shows a single-line drawing of the power circuit for a multiple motor installation. The example shows the procedure to be followed when designing the power circuit.

Power Circuit Conductors

Calculations in this text are based on the assumption that motors are not used for short-time, intermittent, periodic, or varying duty as defined in the electrical code.

For a motor used for any of the above, the branch circuit conductors may have an ampacity at least equal to the full-load current rating of the motor, multiplied by the percentage given in the electrical code for the duty involved.

Check the aforementioned comment in the appropriate electrical code book.

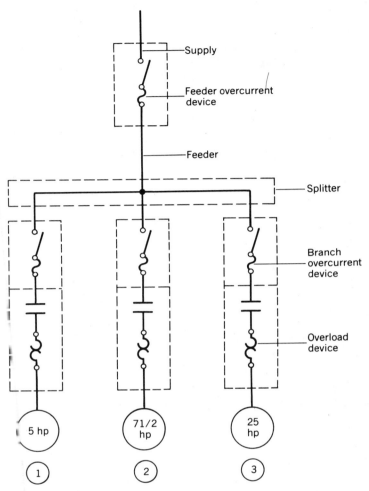

FIGURE 2-2

Single-Line Drawing for the Multiple Motor Installation.

EXAMPLE
A SINGLE MOTOR INSTALLATION

Determine

A. Conductor size.
B. Conduit size.
C. Overcurrent protection.
D. Overload protection.

Data:

Supply voltage	575 V
Motor horsepower	40 hp
Motor type	Squirrel-cage
Service factor	1.15
Phase	Three
Type of overcurrent device	Time-delay fuse
Starting method	Full voltage
Wiring method	R90 X-Link CU in conduit

Solution

Table 2-3 will be used to record the data obtained from the motor installation shown for Figure 2-1.

Step 1 Obtain the estimated full-load amperes (FLA) from the electrical code book.

Answer 41 A

Step 2 The minimum ampacity of the power circuit conductors will be the following.

Answer 125% × 41 A = 51.25 A

Step 3 Using the proper table in the electrical code book, select the correct conductor size.

Answer #6R90 X-Link CU

Step 4 Using the proper table or tables in the electrical code book, select the correct conduit size.

Answer 1 in. (25 mm)

Step 5 To obtain the maximum time-delay fuse, apply the FLA to the table provided.

Answer 80 A fuse (41 A × 175% = 71.75 A)

(The 175% is listed in the code book.)

TABLE 2-3

Chart Recording the Data for the Motor Installation Example, Figure 2-1

HP	FLA	Minimum Ampacity Wire	Wire Size R90 X-Link Cu	Conduit Size	Overload Device (Value)	O/C Time-Delay Fuse
41	41 A	51.25 A	#6	1 in. (25 mm)	51.25 A	80 A

Step 6 To obtain the correct overload heater, apply the FLA (41 A) to the chart on the motor starter cover. The overload value in amperes will be the following.

Answer   41 A × 125% (SF − 1.15) = 51.25 A

EXAMPLE
A MULTIPLE MOTOR INSTALLATION

Determine

A. Conductor sizes.
B. Conduit sizes.
C. Overcurrent protection.
D. Overload protection.

Data:

Supply voltage	575 V
Motor ratings	See Figure 2-2
Motor type	Squirrel-cage
Service factor	1.15
Phase	Three
Type of overcurrent device	Time-delay fuse
Starting method	Full voltage
Wiring method	R90 X-Link CU in conduit

Note: The motors do not start simultaneously.

Solution

Step 1 Full-load current of the motors.
Motor no. 1   6.1 A
Motor no. 2   9 A
Motor no. 3   27 A

Step 2 Minimum ampacity of the branch-circuit conductors
Motor no. 1   FLA × 125% − 6.1 × 125% = (7.625) 15 A
Motor no. 2   FLA × 125% − 9 × 125% = (11.25) 15 A
Motor no. 3   FLA × 125% − 27 × 125% = 33.75 A

Step 3 Minimum conductor size.
        Motor no. 1   <u>#14R90 X-Link CU</u>
        Motor no. 2   <u>#14R90 X-Link CU</u>
        Motor no. 3   <u>#8R90 X-Link CU</u>

Step 4 Minimum conduit size.
        Motor no. 1 $\frac{1}{2}$ in. (12 mm)
        Motor no. 2 $\frac{1}{2}$ in. (12 mm)
        Motor no. 3 $\frac{3}{4}$ in. (19 mm)

Step 5 Maximum branch time-delay fuse.
        Motor no. 1   <u>6.1 A × 175% = (10.67) 15 A</u>
        Motor no. 2   <u>9 A × 175% = (15.75) 20 A</u>
        Motor no. 3   <u>27 A × 175% = (47.25) 50 A</u>

Step 6 Calculated value of the overload device.
        Motor no. 1   <u>6.1 A × 125% = 7.625 A</u>
        Motor no. 2   <u>9 A × 125% = 11.25 A</u>
        Motor no. 3   <u>27 A × 125% = 33.75 A</u>

TABLE 2-4

Chart Recording the Data for the Motor Installation Example, Figure 2-2

	HP	FLA	Minimum Ampacity Wire	Wire Size R90 X-Link CU[a]	Conduit Size	Overload Device (Value)	O/C Time-Delay Fuse
①	5	6.1 A	15 A	#14	$\frac{1}{2}$ in. (12 mm)	7.625 A	15 A
②	7$\frac{1}{2}$	9 A	15 A	#14	$\frac{1}{2}$ in. (12 mm)	11.25 A	20 A
③	25	27 A	33.75 A	#8	$\frac{3}{4}$ in. (19 mm)	33.75 A	50 A

Feeder conductor calculation
(125% × 27) + 9 + 6.1 = 48.85 A   Select #8R90 X-Link CU.[a]

Feeder overcurrent device calculation
50 A + 9 + 6.1 = 65.1 A   Select a 70 A fuse.

Feeder Conduit $\frac{3}{4}$ in. conduit (19 mm)

[a] If an aluminum conductor is required, the size must be selected from the appropriate table in the electrical code book. The conduit must also be larger.

Step 7 To calculate the minimum feeder conductor ampacity we use the following. The feeder conductor ampacity will be 125% of the largest motor, plus the FLA of the remaining motors.
Therefore, (125% × 27) + 9 + 6.1 = 48.85 A. Select #8 R90 X-Link CU.

Step 8 To calculate the feeder overcurrent device, we use the following. The feeder overcurrent device would be the largest branch overcurrent device, plus the FLA of the remaining motors. Therefore, 50 A + 9 + 6.1 = 65.1 A. Select a 70 A fuse.

The above data have been entered on the chart in Table 2-4. The power circuit layout is now complete.

REVIEW EXERCISES

2-1 Is a disconnecting means required within sight of the motor?

2-2 Check the electrical code book to determine what is acceptable as a disconnecting means.

2-3 Is overcurrent protection required to protect the branch power circuit conductors?

2-4 Define a "neutral" conductor as defined by the electrical code.

2-5 Explain "motor service factor."

2-6 What would be the minimum allowable ampacity of the power circuit conductors if the full-load amperes of a motor is 32 A, and the motor is started on full line voltage?

PROBLEM 2A
A SINGLE MOTOR INSTALLATION

Determine

A. Conductor size.

B. Conduit size.

TABLE 2-5

Chart to Record the Data for the Single Motor Installation, Problem 2A

HP	FLA	Minimun Ampacity Wire	Wire Size R90 X-Link Cu	Conduit Size	Overload Device (Value)	O/C Time-Delay Fuse
50						

C. Overcurrent protection.

D. Overload protection.

for an installation as shown in Figure 2-1, using the following data:

Supply voltage	575 V
Motor horsepower	50 hp
Motor type	Squirrel-cage
Service factor	1.15
Phase	Three
Type of overcurrent device	Time-delay fuse
Starting method	Full voltage
Wiring method	R90 X-Link CU in conduit

Enter the answers on Table 2-5. The solution for Problem 2A will be located in the Instructor's Manual.

Solution (enter calculations)

Step 1 _____

_____

_____

Step 2 _____

_____

Step 3 _____

_____

Step 4 _____

Step 5 _____

Step 6 _____

PROBLEM 2B
A MULTIPLE MOTOR INSTALLATION

Determine

A. Conductor sizes.

B. Conduit sizes.

C. Overcurrent protection.

D. Overload protection.

for an installation as shown in Figure 2-3, using the following data:

Supply voltage	575 V
Motor ratings	See Figure 2-3
Motor type	Squirrel-cage
Service factor	1.15
Phase	Three
Type of overcurrent device	Time-delay fuse
Starting method	Full voltage
Wiring method	R90 X-Link CU in conduit

Enter the answers on Table 2-6. The solution for Problem 2B will be located in the Instructor's Manual.

Note: The motors do not start simultaneously.

WORKSHEET FOR PROBLEM 2B
MULTIPLE MOTOR INSTALLATION

Solution

Step 1 _____

       Motor no. 1 _____

       Motor no. 2 _____

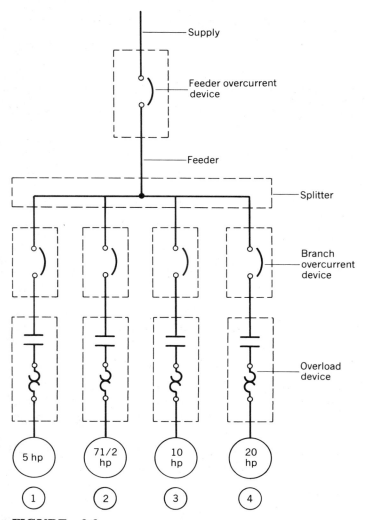

FIGURE 2-3

Single-Line Drawing for the Multiple Motor Installation, Problem 2B.

Motor no. 3 _____

Motor no. 4 _____

Step 2 _____

Motor no. 1 _____

Motor no. 2 _____

TABLE 2-6

Chart to Record the Data for the Multiple Motor Installation, Problem 2B

	HP	FLA	Minimum Ampacity Wire	Wire Size R90 X-Link CU	Conduit Size	Overload Device (Value)	O/C Time-Delay Fuse
①	5						
②	7½						
③	10						
④	20						

Branch Circuits

Feeder conductor calculation (R90 X-Link) _____

Feeder overcurrent device calculation _____

Feeder Conduit _____

Motor no. 3 _____

Motor no. 4 _____

Step 3 _____

Motor no. 1 _____

Motor no. 2 _____

Motor no. 3 _____

Motor no. 4 _____

Step 4 _____

Motor no. 1 _____

Motor no. 2 _____

Motor no. 3 _____

Motor no. 4 _____

Step 5 _____

Motor no. 1 _____

Motor no. 2 _____

Motor no. 3 ⎯⎯⎯⎯⎯⎯⎯⎯⎯⎯⎯⎯⎯⎯⎯⎯

Motor no. 4 ⎯⎯⎯⎯⎯⎯⎯⎯⎯⎯⎯⎯⎯⎯⎯⎯

Step 6 ⎯⎯⎯⎯⎯⎯⎯⎯⎯⎯⎯⎯⎯⎯⎯⎯⎯⎯⎯⎯⎯⎯

Motor no. 1 ⎯⎯⎯⎯⎯⎯⎯⎯⎯⎯⎯⎯⎯⎯⎯⎯

Motor no. 2 ⎯⎯⎯⎯⎯⎯⎯⎯⎯⎯⎯⎯⎯⎯⎯⎯

Motor no. 3 ⎯⎯⎯⎯⎯⎯⎯⎯⎯⎯⎯⎯⎯⎯⎯⎯

Motor no. 4 ⎯⎯⎯⎯⎯⎯⎯⎯⎯⎯⎯⎯⎯⎯⎯⎯

Step 7 ⎯⎯⎯⎯⎯⎯⎯⎯⎯⎯⎯⎯⎯⎯⎯⎯⎯⎯⎯⎯⎯⎯

Step 8 ⎯⎯⎯⎯⎯⎯⎯⎯⎯⎯⎯⎯⎯⎯⎯⎯⎯⎯⎯⎯⎯⎯

Step 9 Feeder conduit. ⎯⎯⎯⎯⎯⎯⎯⎯⎯⎯

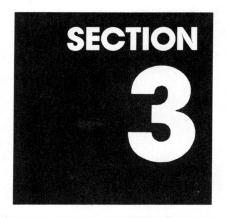

SECTION 3

FULL VOLTAGE STARTING

INTRODUCTION TO
FULL VOLTAGE STARTING

This section will deal with the typical circuits used when motors are started on full line voltage.

The "power supply authority" has definite requirements regarding the starting of large motors. The requirements may vary, but in general the supply authority will insist that the high starting current must not affect the supply system or disturb another consumer's equipment.

When a motor is started by connecting it directly across the line (full voltage), the starting current could be 600% of the nameplate current rating of the motor. If this high in-rush current does not affect the power supply system and the machinery will stand the high starting torque, then full voltage starting may be acceptable.

MANUAL FULL
VOLTAGE MOTOR STARTERS

Manual motor starters may be obtained in single-pole, double-pole, and three-pole designs. The single-pole manual motor starter shown in Figure 3-1 provides one overload heater to protect the motor windings. The double-pole manual motor starter shown in Figure 3-2 also provides one overload heater

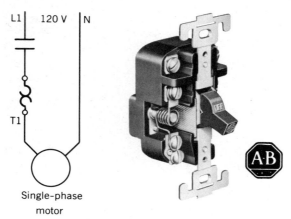

FIGURE 3-1

Single-Phase Motor Connected to a Single-Pole Manual Motor Starter.

to protect the motor windings. The three-pole manual motor starter shown in Figure 3-3 provides three overload heaters to protect the motor windings.

The power circuit contacts of the three manual motor starters shown are unaffected by the loss of voltage, so consequently will remain in the closed position when the supply voltage fails. When the motor is running and the supply volt-

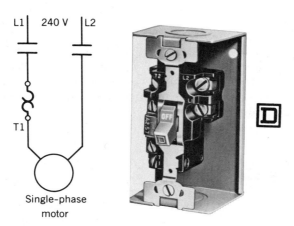

FIGURE 3-2

Single-Phase Motor Connected to a Double-Pole Manual Motor Starter.

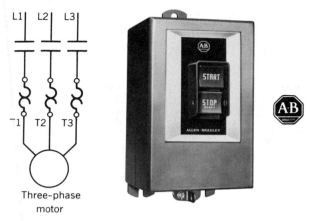

FIGURE 3-3

Three-Phase Motor Connected to Three-Pole Manual Motor Starter.

age fails, the motor will stop and restart automatically when the supply voltage is restored. This will place the three manual motor starters in the classification of "no voltage release."

No voltage release means that the motor will stop when there is a supply voltage failure and restart automatically when the supply voltage is restored.

This term generally refers to a circuit controlled by energizing a holding coil, but the manual motor starters must still be classified as a no voltage release installation.

Magnetic motor control and no voltage release will be dealt with in further detail later in this section.

At this point, check the electrical code to determine if the number of overload devices shown in Figures 3-1, 3-2, and 3-3 are adequate to protect the respective motor windings.

Magnetic Across-the-Line Motor Starters

Unlike the manual motor starter in which the power contacts are closed manually, the magnetic motor starter contacts are closed by energizing a holding coil. This enables the introduction of automatic and remote control of the motor. Example circuits are illustrated in Figures 3-4 and 3-5.

Closing the S' switch will connect the M holding coil across line voltage. The S' switch could be installed at a remote location and still be able to control the M holding coil. A magnetic motor starter may have only one overload contact or

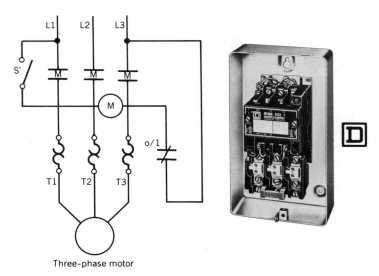

FIGURE 3-4

Wiring Diagram: Power and Control Circuits of a Magnetic Motor Starter Controlled by a Single-Pole Switch.

three overload contacts connected in series. Overload contacts are later illustrated in Figures 3-14 and 3-15.

FIGURE 3-5

Schematic Diagram of the Two-Wire Control Circuit.

MAGNETIC MOTOR STARTER

A magnetic motor starter is an electrically operated switch rated in horsepower and provides for overload protection.

MAGNETIC CONTACTOR

A magnetic contactor is an electrically operated switch rated in amperes. Horsepower ratings are also available for most

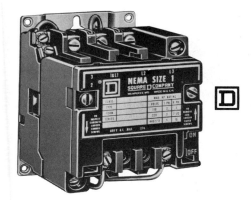

FIGURE   3-6
Magnetic Contactor.

contactors. The magnetic contactors in smaller sizes are referred to as relays. A contactor or relay consists of a set of contacts and a coil. Energizing the coil will open or close the contacts. Contactors are used to interrupt heavy loads, while relays are generally rated at 15 A or less, and are mainly used in control circuits. Control relays are often used to allow low voltage pilot devices to control a higher voltage power circuit. Figure 3-6 is a photograph of a magnetic contactor. Control relays may be referred to as an interlocking relay, jog relay, or a voltage failure relay.

Electromagnetic Control

The contacts of a magnetic motor starter, contactor, or relay are closed by energizing an electromagnet. The electromagnet consists of a coil of wire placed on an iron core. Figure 3-7

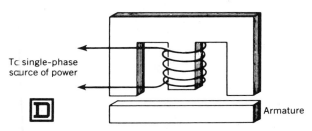

FIGURE   3-7
Simple Electromagnet.

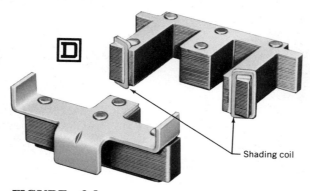

FIGURE 3-8
Magnet Assembly and Armature
Showing the Shading Coils.

illustrates how the coil is placed on the pole piece of an electromagnet.

When the current flows through the coil, the iron of the electromagnet becomes magnetized, attracting the armature. Interrupting the current flow through the coil will cause the armature to drop out due to the air gap in the magnetic circuit.

The contacts are mechanically attached to the armature and will close and/or open as the armature moves.

Shading Coil (Ring)

A shading coil is a single turn of conducting material mounted on the face of the magnet assembly, as depicted in Figure 3-8.

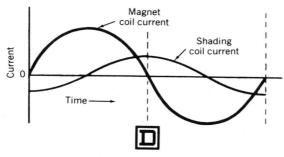

FIGURE 3-9
Magnet Coil Current vs. Shading
Coil Current.

The alternating main magnetic flux induces currents in the shading coil and the currents set up auxiliary flux, which is out of phase from the main flux, and can be easily understood by referring to Figure 3-9.

The auxiliary flux produces a magnetic pull, out of phase from the pull due to the main flux, and this keeps the armature sealed in when the main flux falls to zero. Without the shading coil, the armature would tend to open each time the main flux goes through zero. Excessive noise, wear on the magnet faces, and heat would result.

MOTOR OVERLOAD PROTECTION

The purpose of overload protection is to protect the motor windings from excessive heat, due to motor overloading. The motor windings will not be damaged when overloaded for a short period of time. If the overload should persist, however, the sustained increase in current will cause the overload heater to heat, tripping the mechanism, shutting off the motor.

Types of Overload Devices

The two most common types of overload devices are the *metal alloy* and the *bimetal strip*.

Metal Alloy Overload Device

Metal alloy overload relays, heaters, or elements may be divided into two groups.

Group 1 The first group is one in which the overload heater and the metal alloy (solder pot) are combined in one factory assembled unit, shown in Figure 3-10.

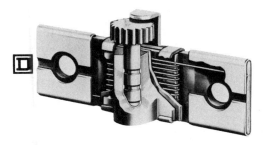

FIGURE   3-10
Metal Alloy Overload Relay.

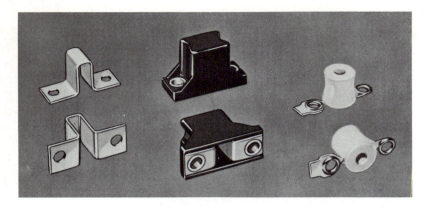

FIGURE   3-11
Metal Alloy Overload Heaters.

Group 2   The second group is one in which the overload heater or element is manufactured separately from the metal alloy (solder pot) and ratchet wheel. The solder pot is factory-installed in the motor starters, and the overload heater is field-installed. Refer to Figure 3-11.

Operation of the Metal Alloy Device

When a motor is loaded beyond the current rating of the overload device, the heater element becomes hot. Heat is transferred to the metal alloy (solder pot), causing the alloy to melt, allowing the ratchet wheel to rotate. The trip mechanism for the manual starters shown in Figures 3-1, 3-2, and 3-3 is mechanical.

In a magnetic motor starter, when the ratchet wheel turns, an overload contact in the control circuit opens, opening the circuit for the electromagnet, stopping the motor, as illustrated in Figure 3-12.

Bimetal Overload Device

Bimetal overload protection consists of an overload heater element placed in the motor starter adjacent to a bimetal strip.

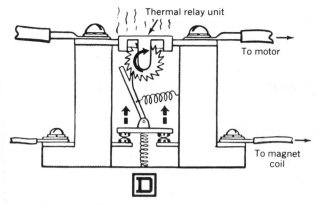

FIGURE 3-12

Operation of melting alloy overload relay. As heat melts alloy, ratchet wheel is free to turn; spring then pushes contacts open.

Operation of the Bimetal Device

The bimetal strip is comprised of two dissimilar metals fused together and secured at one end. When the overload heater becomes hot, due to an overloaded condition, the heat is transferred to the bimetal strip. The metals expand at a different rate, causing the strip to bend in a predetermined direction, mechanically tripping the manual motor starter.

In the magnetic motor starter, an overload contact opens, deenergizing the electromagnet and stopping the motor.

An overload device requires two factors in order to operate: increased current plus time.

Selecting the Correct Overload Device

When selecting an overload relay, heater, or element for *full voltage* starting, obtain the full-load current (FLA) and the service factor stamped on the motor nameplate.

Take the FLA and the service factor to the chart located inside the motor starter and select the correct overload heater by catalogue number. The value in amperes of the heater will be a maximum of

Service factor of 1.0
FLA × 115%
Service factor of 1.15
FLA × 125%

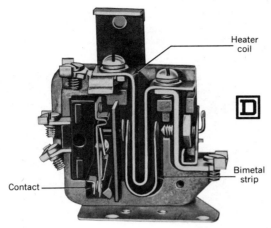

Heater coil

Bimetal strip

Contact

FIGURE 3-13

Bimetallic Overload Relay with Side Cover Removed.

The motor control manufacturers have already calculated these percentages into the overload heater element.

Pay close attention to any special instructions listed on the chart. The percentages used are designated in the electrical code book.

Overload Contacts

In a three-phase magnetic motor starter, you will find three overload heater elements. Some motor control manufacturers use one overload contact, while others use three overload contacts. If one overload contact is used, an overload on any line will cause the overload contact to open. Figure 3-14 illustrates how the system operates.

When three overload contacts are used, the overload contacts are connected in series as shown in Figure 3-15. Each contact is controlled by the adjacent overload heater.

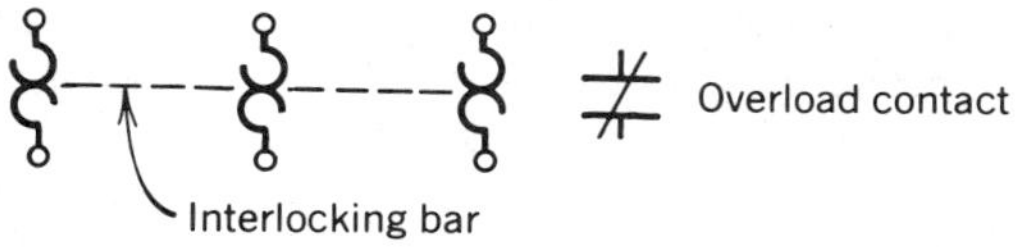

Overload contact

Interlocking bar

FIGURE 3-14

A Single Overload Contact Controlled by Three Overload Heaters.

FIGURE 3-15
Three Overload Contacts Con-
nected in Series.

WIRING DIAGRAM
VERSUS LINE DIAGRAM

A wiring diagram attempts to show the physical location of all
components. Coils, contacts, motors, and the like are shown
in the actual position that would be found on an installation. A
wiring diagram would make it easier to determine the required
number of conductors between points in the circuit, *but* it is
difficult to trace the circuit involved.

A schematic diagram should be used when designing or
troubleshooting an installation. Control components are rearranged to simplify the tracing of the circuit. A tradesperson
must develop the ability to translate a wiring diagram into a
schematic, and a schematic diagram into a wiring diagram.

Compare the wiring diagram in Figure 3-16 with the schematic diagram in Figure 3-17. Apply the comments just made
about wiring and schematic diagrams.

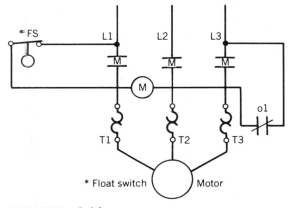

FIGURE 3-16
Wiring Diagram of a No Voltage
Release Circuit.

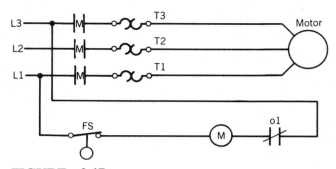

FIGURE 3-17

Schematic Diagram of a No Voltage Release Circuit.

When the starting method is full voltage, the power circuit is often assumed and omitted from the schematic diagram. If the power circuit was removed from Figure 3-16 and the control circuit left untouched, the circuit would resemble Figure 3-18. If the circuit in Figure 3-18 was straightened, the control circuit would resemble Figure 3-19. If the power circuit was removed from Figure 3-20 and the control circuit left un-

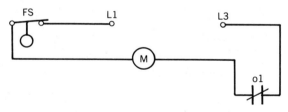

FIGURE 3-18

Control Circuit Shown in Figure 3-16 with the Power Circuit Removed.

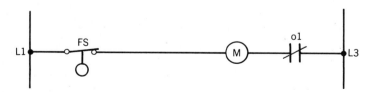

FIGURE 3-19

Control Circuit Shown in Figure 3-18 Straightened to Form a Line Diagram.

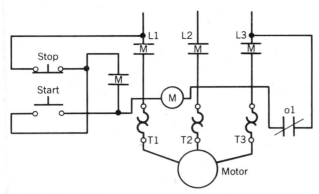

FIGURE 3-20

Wiring Diagram of a No Voltage Protection Circuit.

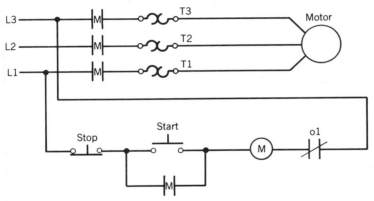

FIGURE 3-21

Schematic Diagram of a No Voltage Protection Circuit.

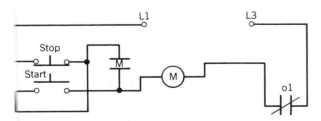

FIGURE 3-22

Control Circuit Shown in Figure 3-20 with the Power Circuit Removed.

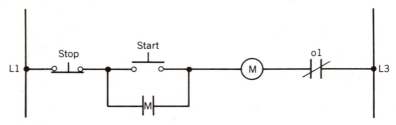

FIGURE 3-23

Control Circuit Shown in Figure 3-22 Straightened to Form a Line Diagram.

touched, the circuit would resemble Figure 3-22. If the control circuit in Figure 3-22 was straightened, the control circuit would resemble Figure 3-23.

Transposing Wiring Diagrams to Line Diagrams and Vice Versa

When drawing a control circuit, attempt to use the line diagram method. On completion, the line diagram must be transferred to a wiring diagram. This operation must be done carefully. It is good practice when transposing from one style of drawing to another to transfer one conductor at a time, checking each off as the transfer is made. The terms ladder, schematic, line, and elementary are generally accepted as having the same meaning.

NO VOLTAGE RELEASE VERSUS NO VOLTAGE PROTECTION

When designing motor control circuits or troubleshooting equipment, two major terms must be fully understood:

A. No voltage release.
B. No voltage protection.

In the previous term, the word "no" is often replaced by the word "low" or "under." Technically, the three terms are not the same, but their effect is similar. For example, if an electromagnetic coil requires a given voltage to close a set of contacts and the voltage should be removed (no), the contacts would open. If for some reason the voltage should be reduced (low, under), the contacts will also open. Thus, we see the reason for using the three terms.

No Voltage Release

This term means that the motor will stop when there is a supply voltage failure, and the motor *will* restart automatically when the supply voltage is restored. The pilot device is unaffected by the loss of voltage and its contacts will remain closed. The term "two-wire" control is given to this type of circuit. Pilot devices such as float switches, limit switches, and temperature controls are single contact devices requiring two conductors.

Circuit Operation, Figures 3-24 and 3-25

When the power supply is established for L1, L2, and L3 and the temperature control is closed, a circuit is completed for the M coil. The M coil becomes an electromagnetic, closing the three M power contacts. If an overload should occur, the overload contact will open, deenergizing the M coil, opening the M power contacts, and stopping the motor. Pressing the overload reset button will reestablish the circuit for the M coil, starting the motor.

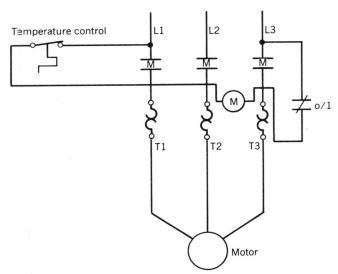

FIGURE   3-24
Wiring Diagram: Two-Wire Control.

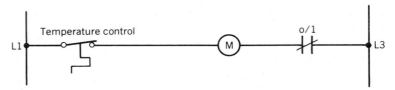

FIGURE 3-25
Schematic Diagram: Two-Wire Control.

This no voltage release circuit would be acceptable for any piece of equipment that is automatically controlled.

The following terms are generally accepted as having the same meaning:

A. Under voltage release (UVR).
B. Low voltage release (LVR).
C. No voltage release (NVR).

No Voltage Protection

This term means that the motor will stop when there is a supply voltage failure, and the motor *will not* restart automatically when the supply voltage is restored.

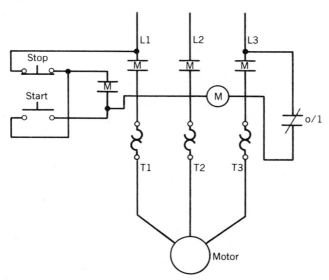

FIGURE 3-26
Wiring Diagram: Three-Wire Control.

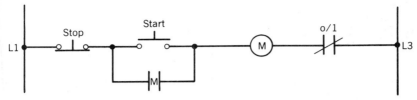

FIGURE 3-27
Schematic Diagram: Three-Wire Control.

Circuit Operation, Figures 3-26 and 3-27

Pressing the start button completes a circuit for the M coil, which closes the M power contacts, starting the motor. At the same instant, the M contact that is connected in parallel with the start button closes and maintains the circuit for the electromagnet when the start pushbutton is released. When the supply voltage fails, the electromagnet is deenergized, opening the M contacts and stopping the motor. The control circuit is now in an open state. On resumption of the supply voltage, the motor will not start automatically.

Any piece of equipment that could injure a person when restarting automatically must be equipped with a no voltage protection circuit. Check this comment in the electrical code book.

The following terms are generally accepted as having the same meaning:

A. Under voltage protection (UVP).
B. Low voltage protection (LVP).
C. No voltage protection (NVP).

DEVELOPING A CONTROL CIRCUIT

Many motor control circuits are designed by electrical designers or electrical engineers. The schematic diagram is generally used when designing circuits. When the electrician obtains the schematic, he or she must transpose the circuit into a wiring diagram. Refer to previous comments in this section under "transposing."

Complex control circuits are comprised of multiple two-wire and three-wire circuits. Control circuits should be designed carefully and methodically. We suggest that a physical layout of the various components be prepared. Understand the operation of each part of the circuit—for example, pilot devices, relays, motor starters, and so on.

Fully understand the operation of the equipment or machine. Commence the control circuit with the initiating device—start button, float switch, pressure switch, and the like. From this point on, continue to design the circuit, one step at a time. When the circuit does not perform the correct sequence of operation required, do not proceed. Locate the error before continuing. Keep in mind

A. Contacts open or close to deenergize or energize coils.

B. Coils are energized or deenergized to open or close contacts.

MOTOR CONTROL CIRCUITS

The following text will focus on various circuits used for full voltage starting. The power circuit will be omitted from these circuits. Figure 3-26 will provide a power circuit if a reference is required.

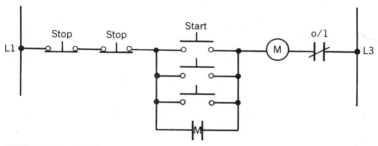

FIGURE   3-28
Multiple Start-Stop Pushbutton Control Circuit. Note: A) The stop buttons are connected in series. B) The start buttons are connected in parallel.

HAND-OFF-AUTO CONTROL CIRCUIT

The hand-off-automatic control circuit (HOA) shown in Figure 3-29 uses a three-position selector switch. Placing the selector switch in the *hand* position completes the circuit for the M coil, energizing the M coil, starting the motor. The *off* position maintains an open circuit for the M coil. When the

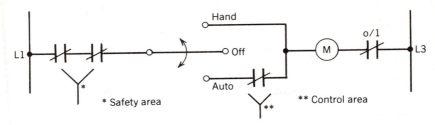

FIGURE 3-29

Hand-Off-Auto Control Circuit.

selector switch is placed in the *auto* position, a circuit is completed for the M coil if the two-wire control is closed. If the two-wire control is in the open position and operation of the equipment is desired, simply place the selector switch in the hand position. Care must be taken to ensure that the selector switch is returned to the automatic position.

The selector switch must never be allowed to bypass the safety controls. The safety controls would be any device that makes the equipment safe to operate.

Example: The safety controls on a boiler may be a low water cut-off, high pressure cut-off, high temperature cut-off, and so on.

This is a no voltage release circuit.

INDICATING PILOT LIGHT CIRCUITS

Figure 3-30 shows two indicating pilot lights. Lamp no. 1 will indicate when the motor is running, while lamp no. 2 will indicate when the motor is not running.

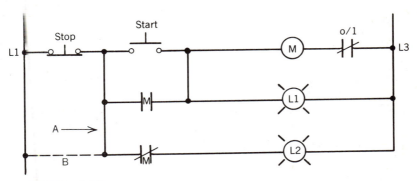

FIGURE 3-30

Pilot Light Control Circuit.

Too much trust is often placed in pilot lights to indicate a particular situation. Lamp bulbs are not always maintained in proper working condition. If it is of the utmost importance to know when the motor is in or out of operation, two lamps should be installed for each condition (on, off). The maintenance staff must then monitor the condition of the pilot lights. The stop button may be locked in the open position. If this happened, for example, lamp no. 2 would not indicate that the motor is not operating. Jumper A must be removed and jumper B installed. The M N/C contact may be added to most magnetic motor starters.

JOGGING

Jogging is defined as an operation in which the motor runs when a pushbutton is pressed and will stop when a pushbutton is released. Jogging is used on machinery in which the motor must run for short periods of time to allow machine setup. Another term for jogging is inching.

Figure 3-31 is a simple circuit that incorporates a single-pole selector switch. When the selector switch is placed in the *run* position, the maintaining circuit is not broken. If we press the start button, the M coil circuit is completed and maintained. Turning the selector switch to the jog position opens the maintaining circuit. Pressing the start button completes the circuit for the M coil, but the maintaining circuit is open. When the start button is released, the M coil is deenergized. The start button doubles as a jog button.

Figure 3-32 shows a jog circuit requiring a double contact jog pushbutton: one N/C contact and one N/O contact. The use of this circuit is widespread and has been so for many years.

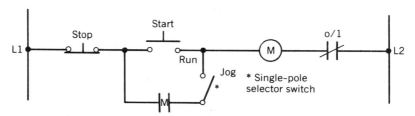

FIGURE 3-31
Selector Switch Jog Control Circuit.

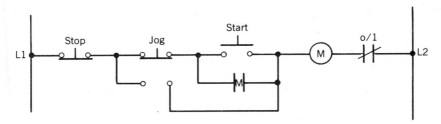

FIGURE 3-32

Jog-Start-Stop Pushbutton Control Circuit.

Pressing the start pushbutton completes a circuit for the M coil, causing the motor to start, and the M contact maintains the M coil circuit.

With the M coil deenergized and the jog pushbutton then pressed, a circuit is completed for the M coil around the M contact. The M contact closes, but the maintaining circuit is incomplete, as the N/C jog button is open.

This control circuit is hazardous. What is the reason for this? The N/C jog button could reclose before the maintaining contact opens, causing the M coil to remain energized. The operator would expect the motor to stop, and it may not.

The control circuit shown in Figure 3-33 is much safer than the previous circuit. A single contact jog pushbutton is used, plus the circuit incorporates a *jog relay* (control relay).

Pressing the start pushbutton completes a circuit for the CR coil, closing the CR(1) and CR(2) contacts. CR(1) contact completes the circuit for the M coil, starting the motor. The M maintaining contact closes, maintaining the circuit for the M

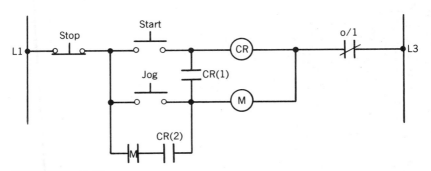

FIGURE 3-33

Jog-Start-Stop Pushbutton Control Circuit Using a Jog Relay.

coil. The motor continues to run. Pressing the stop button stops the motor.

Pressing the jog button energizes the M coil only, starting the motor. Both CR contacts remain open and the CR coil deenergized. The M coil will not remain energized when the jog pushbutton is released. The coil sequence is the following:

- Press the start pushbutton: CR–IN
                                                    M–IN
- Press the jog pushbutton: M–IN

REVERSING DIRECTION OF ROTATION OF A THREE-PHASE MOTOR

To reverse the direction of a three-phase motor, simply interchange any two phase conductors at the motor or the motor starter. If the reversing of the motor is a required operation of the machine, then reversing motor starters are available.

Manual Three-Phase Reversing Motor Starters

Care must be taken when selecting a manual reversing motor starter. If remote control is required, a manual reversing motor starter is not practical. Keep in mind that a manual, maintained contact reversing starter is a no voltage release circuit. Their use would be unacceptable if a requirement for the machine is a no voltage protection circuit. The power circuit for a manual reversing motor starter is shown in Figure 3-34.

Figure 3-34 indicates that the forward (F) contacts close together and the reverse (R) contacts close together. If both sets of contacts were allowed to close at the same time, a short circuit would develop. The two sets of contacts are mechanically interlocked to prevent this from happening. Figure 3-34 identifies the T terminals for both forward and reverse. Two sets of overload devices are required for such motor starters.

Reversing Magnetic Motor Starter

No voltage release, no voltage protection, automatic control, and remote control are all features available by using a reversing magnetic motor starter. The power circuit of a three-phase

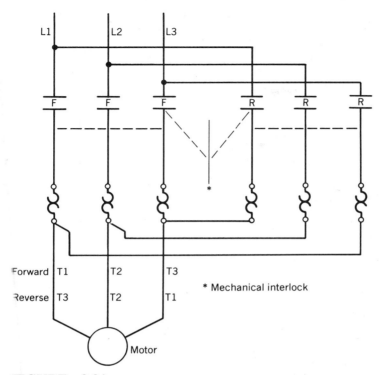

FIGURE 3-34

Three-Phase Manual Reversing Motor Starter Power Circuit.

reversing magnetic motor starter is shown in Figure 3-35. One set of overload devices is required in a reversing magnetic motor starter.

Closing the F power contacts causes the motor to rotate in the forward rotation. The R power contacts are not able to close while the F contacts are closed. A mechanism called a mechanical interlock is an integral part of all reversing motor starters. The mechanism varies in construction and appearance from one manufacturer to another, but its purpose is the same: to prevent both sets of power contacts from closing at the same time.

Figure 3-36 shows a schematic diagram of a control circuit to control the power circuit in Figure 3-35. It also illustrates the control circuit generally used for size 00 reversing magnetic motor starters.

In operation, pressing the forward pushbutton completes the circuit for the F coil, closing the F maintaining contact,

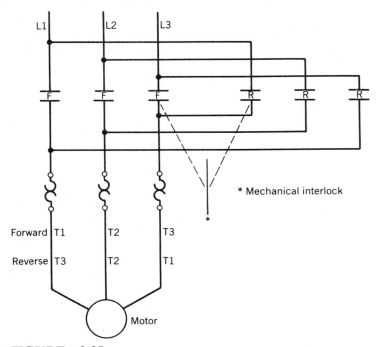

FIGURE   3-35

Three-Phase Magnetic Reversing Motor Starter Power Circuit.

sealing the circuit for the F coil. The F power contacts have closed and the motor is running.

Pressing the reverse pushbutton completes the circuit for the R coil. The R armature will attempt to close the R contacts, but the mechanical interlock will prevent this from happening.

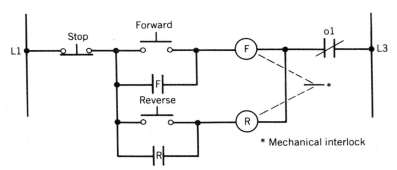

FIGURE   3-36

Forward-Reverse-Stop Control Circuit with Mechanical Interlock.

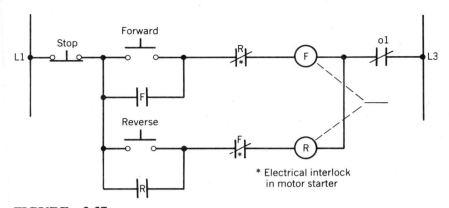

FIGURE 3-37

Forward-Reverse-Stop Control Circuit with Electrical Interlock in the Motor Starter.

In sizes 0 and larger, the control circuit shown in Figure 3-37 is generally used to control the power circuit in Figure 3-35. In addition to the mechanical interlock, the larger reversing magnetic motor starters are equipped with electrical interlocks.

In operation, pressing the forward pushbutton completes the circuit for the F coil, closing the N/O F maintaining contact and sealing the circuit for the F coil. At the same instant, the N/C F contact opens. The F power contacts have closed and the motor is running.

Pressing the reverse pushbutton will not complete the circuit for the R coil, as the N/C F contact is now open. This control circuit is used so that dependence on the mechanical interlock is not needed.

To reverse the motor with this control circuit, the operator must press the stop button to deenergize the respective coil, reclosing the respective N/C contact.

Figure 3-38 is an example of a control circuit that will allow the selecting of rotation, without pressing the stop button. Care should be exercised in selecting this circuit.

In operation, pressing the forward pushbutton completes the circuit for the F coil, closing the N/O F maintaining contact, and opening the N/C F electrical interlock. The F power contacts have closed and the motor is running.

Pressing the reverse pushbutton deenergizes the F coil and completes the circuit for the R coil. The motor reverses rotation immediately. The driven machinery and the high currents involved must be considered before using this control circuit.

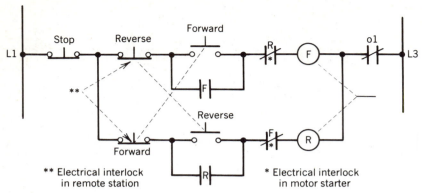

FIGURE 3-38

Forward-Reverse-Stop Control Circuit with Electrical Interlocks in the Remote Pushbutton and the Motor Starter.

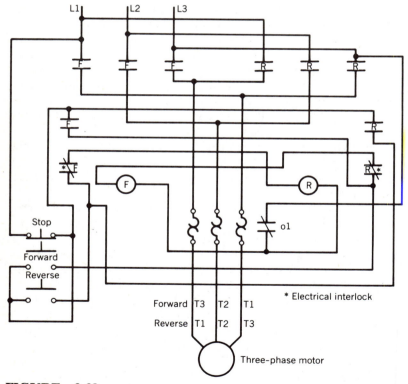

FIGURE 3-39

Wiring Diagram Combining Figures 3-35 and 3-37.

The wiring diagram in Figure 3-39 combines the power circuit of Figure 3-35 and the control circuit of Figure 3-37. The schematic shown in Figure 3-37 was transferred one conductor at a time. If we compare the schematic with the wiring diagram, it should be emphasized that the transfer must be done very carefully. If care is not exercised when transferring a schematic diagram to a wiring diagram, an error is possible that could render the circuit inoperable.

REVERSING SINGLE-PHASE MOTORS

When reversing a split-phase motor or a single voltage capacitor-start motor, the direction of rotation is reversed by inter-

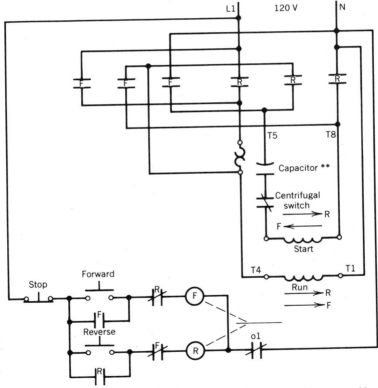

** With the capacitor eliminated from the starting winding, the circuit would now resemble the circuit diagram for a split-phase motor, controlled by a reversing magnetic motor starter.

FIGURE 3-40

The Power and Control Circuits for a Single Voltage Capacitor-Start Motor Connected to a Reversing Magnetic Motor Starter.

changing the starting winding leads, or the running winding leads.

When reversing a dual voltage capacitor-start motor, interchange the starting winding leads or reconnect the T5 lead as shown in the power circuit diagram of Figure 3-41.

If remote control is required to start and reverse the direction of rotation of a split-phase or capacitor-start motor, a reversing magnetic motor starter may be used, in conjunction with a forward–reverse–stop pushbutton station.

Reversing Magnetic Motor Starters to Control Single-Phase Motors

The most common single-phase motors, which may be controlled by a reversing magnetic motor starter, are the split-phase and capacitor-start motors.

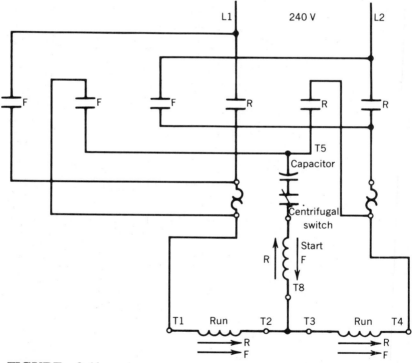

FIGURE   3-41

The Power Circuit for a Dual Voltage Capacitor-Start Motor Connected to a Reversing Magnetic Motor Starter.

Unlike the three-phase motor, the split-phase and the capacitor-start motors must be allowed to slow down before attempting to reverse the direction of rotation. In each of the above single-phase motors, the centrifugal switch in the starting winding circuit opens at approximately 75% of the motor speed and must be allowed to reclose before the motor will reverse. This would then eliminate the control circuit shown in Figure 3-38.

Instantaneous Current Flow

Unlike DC circuits, in which the direction of the current flow can be established, AC current flow alternates. To indicate current flow in Figures 3-40 and 3-41, we determined that the instantaneous current flow direction occurs from L1 to N (neutral), and L1 to L2. Changing the instantaneous current flow direction as indicated will reverse the direction of rotation of the motors shown.

REVIEW EXERCISES

3-1   Define "full voltage starting."
3-2   Explain the function of a magnetic motor starter.
3-3   Define "a magnetic motor starter."
3-4   Define "a magnetic contactor."
3-5   What is the function of the shading ring installed in an AC magnetic motor starter, or magnetic contactor?
3-6   What is the purpose for installing an overload device in a motor circuit?
3-7   Explain the operation of an overload contact.
3-8   Explain "no voltage release."
3-9   Explain "no voltage protection."
3-10 How are stop buttons connected for two-point control?
3-11 Does the "hand-off-auto control circuit" provide "no voltage release" or "no voltage protection"?
3-12 Define the term "jogging."

3-13 How is the direction of rotation of a three-phase motor reversed?

3-14 How is the direction of rotation of a split-phase motor reversed?

3-15 Explain the operation of the mechanical interlock in a reversing magnetic motor starter.

SECTION 4

SPECIALTY CIRCUITS

INTRODUCTION TO SPECIALTY CIRCUITS

Section 4 will be devoted to specialty circuits, involving inter-locking, preferred loads, possible hazards, and the require-ments of the electrical code.

CONTROL TRANSFORMER

Step-down control transformers are installed when the control circuit components are not rated for line voltage. The primary side of the control transformer will be line voltage, while the secondary voltage will be the voltage required for the control components.

Figure 4-1 shows the physical layout of a full voltage mag-netic motor starter, controlled by a start–stop remote station. The power supply will be 575 V, three-phase, and the control circuit voltage will be 120 V.

The point to be discussed is the jumper A, which connects X2 of the control transformer secondary winding to ground.

Some inspection authorities do not require grounding of the secondary winding as shown in Figure 4-2. This point should be checked in the electrical code for the district or area in-volved. However, many electrical designers, engineers, and

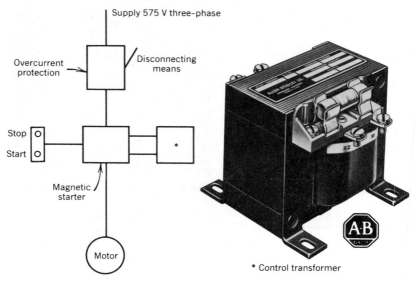

FIGURE 4-1

Physical Layout for the Power and Control Circuits for a Single Motor Installation, with a Control Transformer to Provide Low Voltage for the Control Circuit.

motor control manufacturers recommend, if not insist, that the secondary winding be grounded as shown.

Take time to analyze the circuit, and you will be convinced that this is the correct installation. The electrical code states, paraphrased

If a control circuit is supplied from a grounded circuit, it must be so connected, that an accidental ground in the control circuit will not start the motor, or make the stop button or control inoperative.

Figure 4-3 shows the control transformer properly grounded. Brackets have been placed around the part of the control circuit that is field-wired and possibly remote from the magnetic motor starter.

A strong case must be made in favor of grounding the secondary winding of the control transformer as shown in Figure 4-3.

Figure 4-4 is a photo of a start–stop pushbutton supplied by a piece of armored cable.

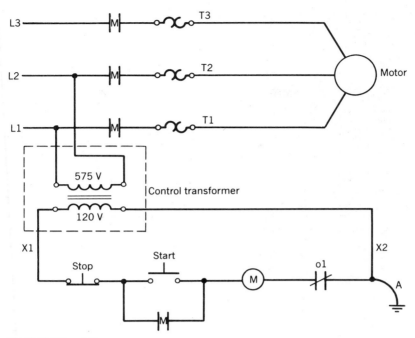

FIGURE   4-2
Schematic Diagram for the Installation Shown in Figure 4-1.

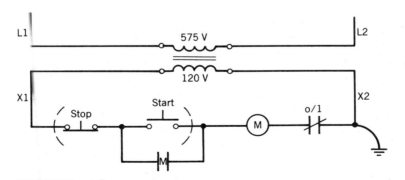

FIGURE   4-3
Schematic Diagram Showing the Correct Grounding for the Secondary Winding of a Control Transformer.

Fault B

Armored ← cable

FIGURE 4-4

**Start–Stop Pushbutton Station
Showing an Accidental Ground
at Point B.**

If vibration caused the insulation on the control circuit conductors to fray, thereby making the conductor touch ground at point B, Figure 4-5, the transformer would be short-circuited, deenergizing the M coil, stopping the motor. That is what should happen. If a fuse had been installed in the secondary winding circuit, the transformer winding would not be damaged. Now consider the same fault B, but with the secondary winding grounded at point C (Figure 4-6).

When the fault at B occurred, the circuit for the M coil was completed through the grounded cable, starting the motor unexpectedly. This would be extremely hazardous. Pressing the stop button will not deenergize the M coil.

If the control circuit was a two-wire temperature, pressure, or liquid level control, the same hazard is present. (Figure 4-7).

When connecting a control circuit to the secondary winding

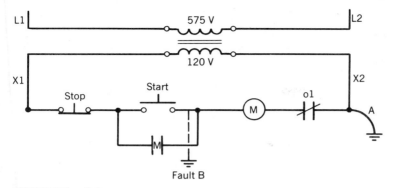

FIGURE 4-5

Schematic Diagram Showing the Correct Grounding at Point A and an Accidental Ground at Point B.

cf a control transformer, check the secondary grounding and connect accordingly.

INTERLOCKING

Interlocking simply means to connect together pieces of electrical equipment. The circuits, whether power or control, are said to be interlocked when one circuit controls another circuit. All interlocking must be done in a safe manner. Many times interlocking will involve multiple circuits. The safety of the electrician or service personnel is of utmost importance.

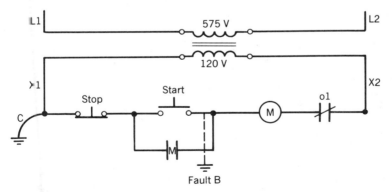

FIGURE 4-6

Schematic Diagram Showing Incorrect Grounding at Point C and an Accidental Ground at Point B.

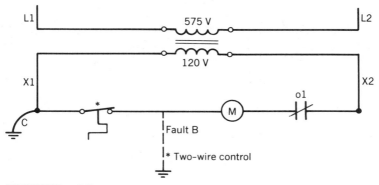

FIGURE   4-7

Schematic Diagram of a Two-Wire Control Circuit Showing Incorrect Grounding at Point C and an Accidental Ground at Point B.

Interlocking Electrical Equipment

Figure 4-8 shows an installation in which the first motor must start before the second.

The broken line in Figure 4-8 indicates the wiring between the two magnetic motor starters. This arrangement is referred to as preferred load. The point is that the interlocking of the two circuits must be done safely.

The electrical code dictates that when the disconnecting means is turned off (e.g., the M1 circuit), all contacts in the M1 motor starter must be dead.

Figure 4-9 shows how the control circuits might be installed. When analyzing these circuits, the hazard involved should be noted. Combine Figures 4-8 and 4-9. If the M1 circuit were disconnected, the M1 contact (*) in the M2 circuit would still be a live contact. Working in the M1 motor starter would be dangerous.

Some inspection authorities may permit the use of an identified shroud (cover) over the M1 (*) contact. The safest installation would be a dual voltage relay.

Dual Voltage Relay (DVR)

A dual voltage relay could be called a dual source relay. Interlocking the same voltages from a dual source still requires a dual voltage relay.

A dual voltage relay consists of an electromagnetic coil and contacts. The coil and contacts are housed in a metal enclo-

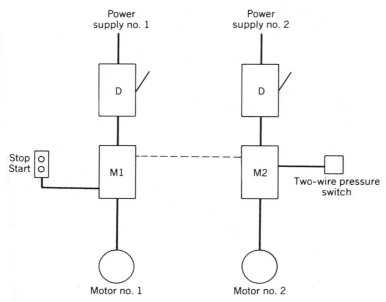

FIGURE 4-8

Physical Layout Showing How Two Motor Installations Might Be Interlocked.

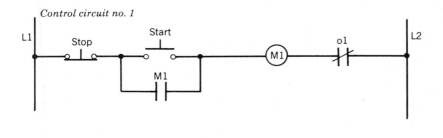

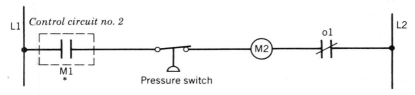

FIGURE 4-9

Schematic Diagram for the Installation Shown in Figure 4-8.

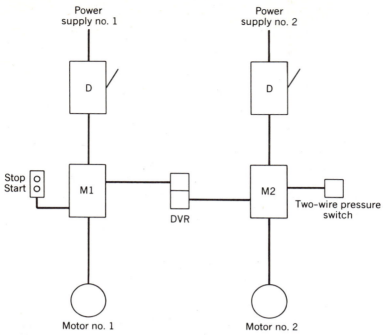

FIGURE 4-10

Physical Layout Showing How Two Motor Installations Should Be Safely Interlocked, Using a Dual Voltage Relay.

sure and are separated by a fixed partition, or divider. Some dual voltage relays have two separate covers, while others have one cover, with two separate inner covers.

The coil may be connected to one supply voltage, while the contact may be connected in another circuit. The divider must never be pierced to allow conductors to pass from one side to the other.

To improve on Figure 4-8, the installation in Figure 4-10 should be considered. The control circuit for the installation in this drawing would be designed as shown in Figure 4-11.

Think safety when designing control circuits, especially when interlocking multiple circuits.

SEQUENCE CONTROL

Sequence control means that the motors must start one after the other in a predetermined order. A good example would be

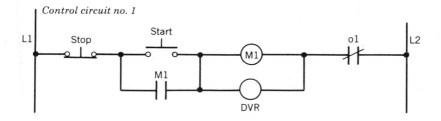

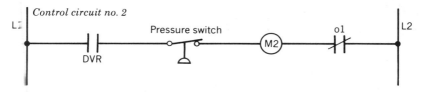

FIGURE 4-11

Schematic Diagram for the Installation Shown in Figure 4-10.

a conveyor system composed of four sections. The sections must start in the correct sequence. If, for some reason, one motor will not start, the next motor in line will not start.

Will the electrical code permit such an installation, as shown in Figure 4-14? Four $\frac{1}{4}$ hp motors are grouped on a single fused disconnecting means with the overcurrent device set at 15A. The answer to the question is *yes*. With this installation, unsafe interlocking or the requirement for dual voltage relays are not factors to be addressed.

The control circuit in Figure 4-15 will show that the interlocking has been done properly.

Sequence Starting of Multiple Motors

Another class of preferred load is called sequence starting. See Figure 4-12.

Figure 4-13 shows the schematic for the control circuit for the sequence control installation in Figure 4-12. The wiring diagram of the power and control circuits for the sequence control installation in Figure 4-12 is illustrated in Figure 4-14.

Note that each magnetic motor starter will provide overload protection, sized to protect each individual motor.

When designing a control circuit, know what is to happen, and then make it happen.

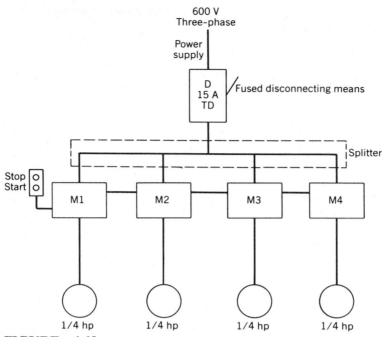

FIGURE 4-12

Physical Layout of a Multiple Motor Installation for Sequence Starting.

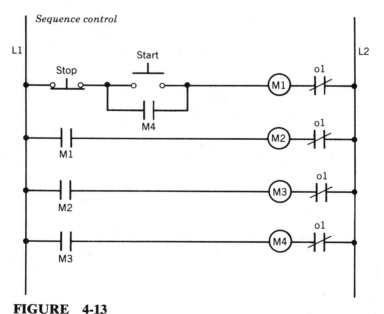

FIGURE 4-13

Schematic Diagram for the Installation Shown in Figure 4-12 (control circuit).

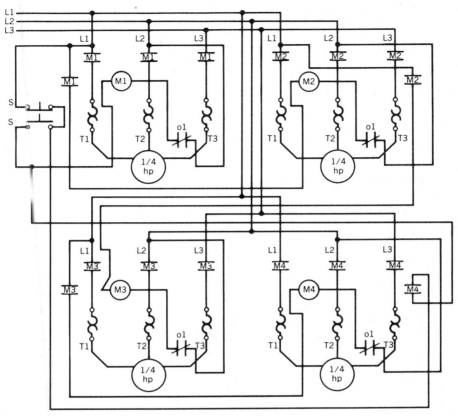

FIGURE 4-14

Wiring Diagram for the Installation Shown in Figure 4-12 (power and control circuits).

TRANSPOSING CIRCUITS

Compare the ladder diagram in Figure 4-13 and the wiring diagram in Figure 4-14. The circuit should be designed using the schematic or ladder diagram. Transposing the schematic diagram to the wiring diagram can be a bit tricky. An error may be made when transposing the circuit, and it will possibly be difficult to locate on a wiring diagram such as Figure 4-14.

Transpose the control circuit in Figure 4-13 to the power circuit very carefully, one conductor at a time. Develop the attitude that when a circuit is thoughtfully designed, works correctly on paper, and is properly installed, it *will* work.

TROUBLESHOOTING

Time spent studying the control circuit in Figure 4-13 will not be wasted. For example, if the M1 N/O contact is broken and will not close, and we press the start button, what will happen? The M1 coil will be energized; only the M1 motor will start. Release the start button and the M1 motor will stop.

Take time to fault the circuit in Figure 4-13 and determine the outcome. This will be your introduction to troubleshooting motor control installations.

Circuit Operation, Figure 4-13

If we press the start button, a circuit is completed for the M1 coil, closing the M1 contact, energizing the M2 coil, closing the M2 contact, energizing the M3 coil, closing the M3 contact, energizing the M4 coil, closing the M4 contact, sealing the circuit. Pressing the stop button breaks the circuit for the M1 coil, deenergizing the coils in the same sequence. This is sequence-stopping as well as sequence-starting. If an overload contact should open, the motors will stop in sequence. For example, if the overload contact for the M3 motor should open, the stopping sequence would be M3, M4, M1, and then M2.

Every installation must be designed to meet the needs and requirements of the particular machine involved.

RING CONTROL CIRCUIT

In addition to being a sequence control, the circuit in Figure 4-13 is a ring circuit. A ring circuit is one in which the motors start in sequence, and the maintaining contact of the last magnetic motor starter seals the circuit.

PLUGGING

Plugging is a method of stopping a polyphase motor quickly, by momentarily connecting the motor for the reverse rotation, when the motor is running. Plugging a motor more than five times a minute requires a larger motor starter than is required for a given motor.

Figure 4-15 is a photo of a *zero-speed switch*. This switch is often referred to as a *plugging* switch or an *anti-plugging*

FIGURE 4-15
Zero-Speed Switches.

switch. The correct name, however, is a zero-speed switch; the other terms are used to denote the intended use.

If the machine, because of the load, should not be stopped abruptly, it should not be plugged to a stop.

Installing a Zero-Speed Switch

The zero-speed switch is coupled to a moving shaft on the machinery whose motor is to be plugged. Each zero-speed switch has a speed range stamped on the nameplate; for example, 50 to 200 rpm. Attempt to operate the zero-speed switch within the limits of the rated speed range.

Operation

As the zero-speed switch rotates, centrifugal force or a magnetic clutch causes the contacts to open or close, depending on the intended use.

Symbols

The symbols for the zero-speed switch are shown in Figure 4-16 illustrating the intended use.

Plugging a Motor from One Direction Only

One Method

Figure 4-17 illustrates the method of connecting the components for the circuit shown in Figure 4-18.

Circuit Operation, Figure 4-18

Pressing the start button energizes the F coil, causing the motor to rotate in the forward rotation. The N/C F contact opens, while the forward rotation causes the zero-speed

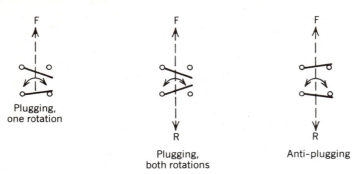

Plugging,
one rotation

Plugging,
both rotations

Anti-plugging

FIGURE 4-16

Contact Positions of the Zero-Speed Switch.

switch contact to close. Pressing the stop button deenergizes
the F coil, closing the N/C F contact and energizing the R coil.
The R power contacts close, causing the motor terminals T1
and T2 to change, reversing the rotation of the motor. The
motor speed decreases rapidly (depending on the load on the

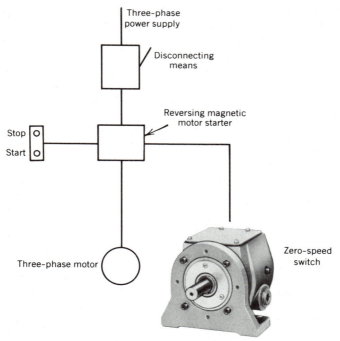

Three-phase
power supply

Disconnecting
means

Reversing magnetic
motor starter

Stop

Start

Three-phase motor

Zero-speed
switch

FIGURE 4-17

**Physical Layout for Plugging a Motor from One Rotation Only
(method one).**

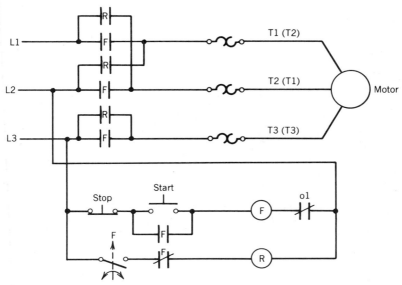

FIGURE 4-18

Schematic Diagram for the Installation Shown in Figure 4-17.

motor) and when the motor reaches near zero rpm, the zero-speed switch contact opens, deenergizing the R coil, preventing the motor from running in reverse.

This particular installation has a built-in hazard. All personnel involved with the servicing of rotating equipment should understand the safe method of checking such machinery. Before attempting to examine any piece of electrically operated equipment, it is imperative that the disconnecting means be locked in the off position. If a person ignored or neglected the correct safety procedure and did not disconnect the power supply, and he or she then pulled the belt on the machine in the forward rotation, it is very possible that the contacts of the zero-speed switch would close, energizing the R coil momentarily, and causing the motor to rotate enough to cause an injury.

Another point is worth mentioning: If an overload contact opened, the motor would still be plugged to a stop. Connecting the R coil to the left side of the overload contact, instead of L2, the circuit would not provide plugging if an overload contact opened.

Second Method

Figure 4-19 illustrates the method of connecting the components for the circuit shown in Figure 4-20.

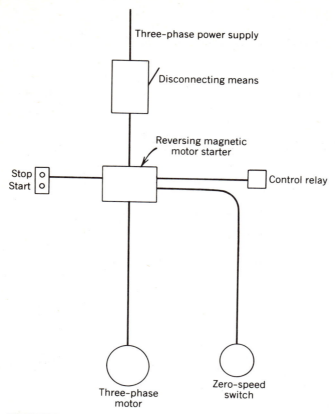

FIGURE 4-19

Physical Layout for Plugging a Motor from One Rotation Only (method two).

Circuit Operation, Figure 4-20

Pressing the start button energizes the F coil, causing the power contacts to close, starting the motor. One N/O F contact maintains the circuit, and the second N/O F contact completes the circuit for the CR control relay coil. At this point, the CR contact is closed and the N/C F contact open. The rotation of the zero-speed switch closes the zero-speed switch contact. Pressing the stop button deenergizes the F coil, closing the N/C F contact and energizing the R coil. The R power contacts close, causing the motor terminals T1 and T2 to change, reversing the rotation of the motor. The motor speed decreases rapidly (depending on the load on the motor) and when the motor reaches near zero speed, the zero-speed

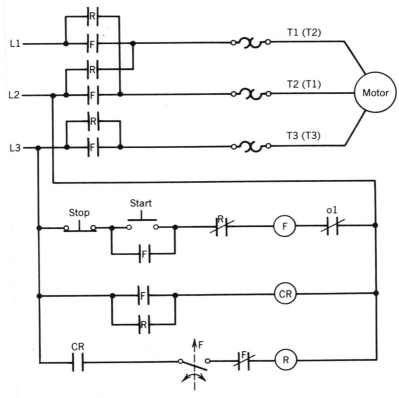

FIGURE 4-20

Schematic Diagram for the Installation Shown in Figure 4-19.

switch contact opens, deenergizing the R coil, and thus preventing the motor from running in reverse.

Closing the zero-speed switch contacts accidentally will not cause the R coil to become energized.

Third Method

Figure 4-21 illustrates the method of connecting the components for the circuit shown in Figure 4-22.

Circuit Operation, Figure 4-22

The second method illustrated in Figures 4-19 and 4-20 has removed the hazard inherent in the circuit shown in Figure 4-18. The installation and circuit in Figures 4-21 and 4-22 also removed the same hazard.

Pressing the start button energizes the F coil, starting the

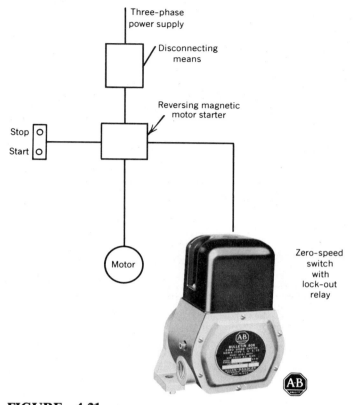

FIGURE 4-21

Physical Layout for Plugging a Motor from One Rotation Only (method three).

motor, and the circuit is maintained by the N/O F contact. The N/C F contact is now open. The LC coil is energized, allowing the zero-speed switch contact to close. The zero-speed switch contact will not close unless the lockout coil is energized.

Pressing the stop button deenergizes the F coil, closing the N/C F contact, and energizing the R coil momentarily. The power contacts close, causing the motor terminals to change, reversing the rotation of the motor. The motor speed decreases rapidly and when the motor reaches near zero speed, the zero-speed switch contact opens, deenergizing the R coil, thereby preventing the motor from running in reverse.

Rotating the zero-speed switch will not allow the motor to start accidentally.

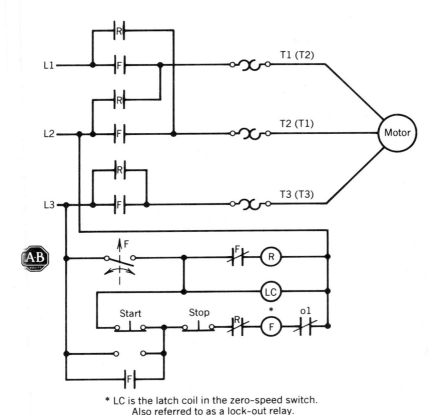

* LC is the latch coil in the zero-speed switch.
Also referred to as a lock-out relay.

FIGURE 4-22

Schematic Diagram for the Installation Shown in Figure 4-21.

Plugging a Motor from Both Directions

Figure 4-23 illustrates the method of connecting the components for the circuit shown in Figure 4-24.

Circuit Operation, Figure 4-24

Compare the location of the overload contacts in the control circuits shown in Figures 4-18, 4-20 and 4-22 with the location of the overload contacts in Figure 4-24. If the overload contact should open in the circuit illustrated in Figure 4-24, it will deenergize the appropriate coil and shut off the motor.

Pressing the forward button opens the R coil circuit and energizes the CR coil. CR(1) contact maintains the circuit for

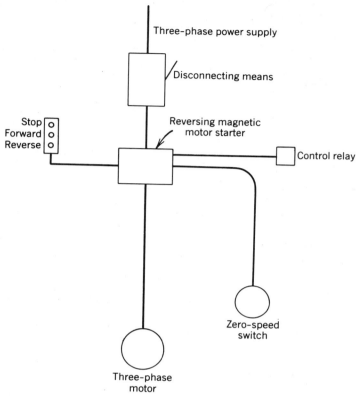

FIGURE 4-23

Physical Layout for Plugging a Motor from Both Rotations.

the CR coil. CR(2) and CR(3) contacts close. The N/C F contact is now open. The zero-speed switch F contact has closed.

Pressing the stop button deenergizes the CR coil, opening the CR(2) and CR(3) contacts, deenergizing the F coil, closing the N/C F contact, and energizing the R coil momentarily. The power contacts close, causing the motor terminals T1 and T2 to change, reversing the rotation of the motor. The motor speed decreases rapidly and when the motor reaches near zero rpm, the zero-speed switch contact opens, deenergizing the R coil, thus preventing the motor from running in reverse.

Rotating the zero-speed switch will not allow the motor to start accidentally.

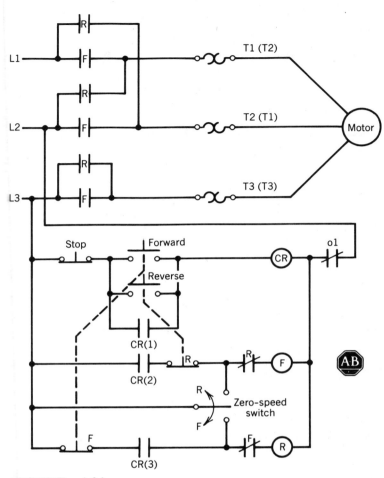

FIGURE 4-24

Schematic Diagram for the Installation Shown in Figure 4-23.

ANTI-PLUGGING

Many machines, large and small, require the motor to be able to reverse, either automatically or manually.

Small machines may not be adversely affected by reversing the motor before coming to a stop. An example of this type of machine would be the lathe. Reversing the motor before allowing it to come to a stop is the normal operation of the lathe.

This same principle is not true on larger pieces of equip-

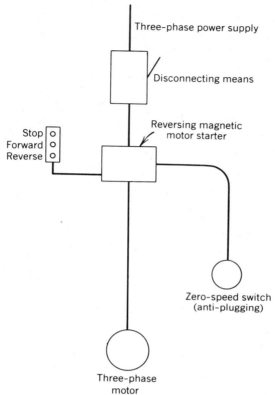

FIGURE 4-25

Physical Layout for Anti-plugging.

ment. The sudden reversing torque applied when a large motor is reversed without slowing the motor speed could damage the driven machinery, plus the extremely high current could affect the distribution system.

There are two methods that may be used to prevent reversing the motor before the motor has slowed to near-zero speed. The first method is to instruct the machine operator to exercise patience and wait for the motor to stop. This method often falls short of perfection. The second method, and by far much more reliable than the first, is to install a zero-speed switch to monitor the equipment and to provide anti-plugging.

Figure 4-25 illustrates the method of connecting the components for the circuit shown in Figure 4-26.

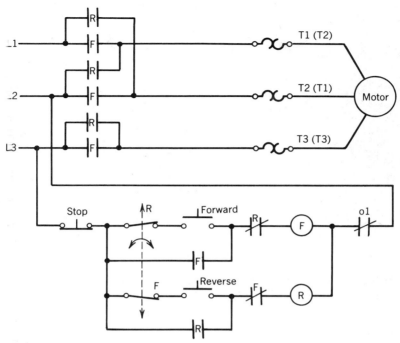

FIGURE 4-26

Schematic Diagram for the Installation Shown in Figure 4-25.

Circuit Operation, Figure 4-26

Pressing the forward button completes a circuit for the F coil, closing the F power contacts, causing the motor to run in the forward rotation. At the same instant, the F N/O maintaining contact closed, maintaining the F coil. The F N/C contact has also opened, providing an electrical interlock in the R coil circuit. The F zero-speed switch contact opens due to the forward rotation of the motor.

Pressing the stop button deenergizes the F coil, disconnecting the motor, and causing the motor to slow down.

Pressing the reverse button will not complete a circuit for the R coil, until the F zero-speed switch contact recloses. When the rotating equipment reaches near-zero speed, the reverse circuit may be energized and the motor will run in the reverse rotation.

This circuit is widely used to prevent damage to equipment.

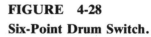

FIGURE 4-27

Six-Point Drum Switch
(cover removed).

FIGURE 4-28

Six-Point Drum Switch.

DRUM SWITCHES

Drum switches are used to control the direction of rotation of
single-phase and polyphase motors. The drum switches
shown in Figures 4-27 and 4-28 have a hp rating and three
positions: reverse, off, and forward. The drum switch does
not provide overcurrent or overload protection. The contacts
are designed to be maintained or momentary and are easily
converted in the field.

Reverse	Off	Forward
1 o— —o 2	1 o o 2	1 o o 2
3 o— —o 4	3 o o 4	3 o o 4
5 o— —o 6	5 o o 6	5 o— —o 6

FIGURE 4-29

Six-Point Drum Switch Contact
Positions.

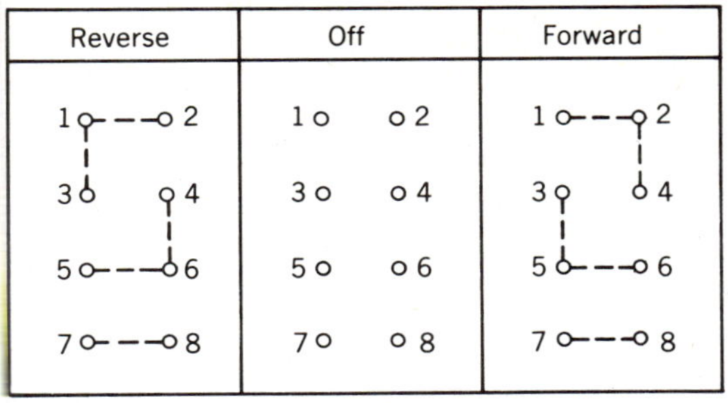

Reverse	Off	Forward
1 o – – o 2 3 o o 4 5 o – – o 6 7 o – – o 8	1 o o 2 3 o o 4 5 o o 6 7 o o 8	1 o – – o 2 3 o o 4 5 o – – o 6 7 o – – o 8

FIGURE 4-30
Eight-Point Drum Switch Contact Positions.

Drum switches are identified by the number of connection points. The three most common drum switches are

A. Six-point.
B. Eight-point.
C. Nine-point.

Figures 4-27 and 4-28 are examples of the six-point drum switch. The internal switching for each of the drum switches is represented in Figures 4-29, 4-30, and 4-31. The contact positions are indicated by dotted lines.

INSTANTANEOUS CURRENT FLOW

Unlike DC circuits, in which the direction of the current flow can be established, AC current flow alternates. In order to indicate current flow in the following drawings, we decided to show instantaneous current flow direction going from L1 to N (neutral), and L1 to L2. Changing the instantaneous current flow direction as illustrated will reverse the direction of rotation of the motors depicted.

Six-Point Drum Switch

Figures 4-32 through 4-37 show the circuit diagrams for various AC motors controlled by six-point drum switches. Arrows have been used to indicate the direction of the instantaneous current flow.

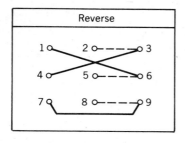

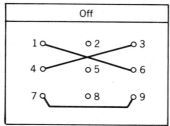

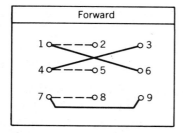

FIGURE 4-31

Nine-Point Drum Switch Contact Positions.

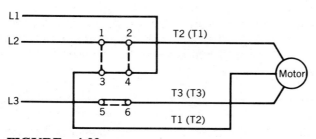

FIGURE 4-32

Three-Phase Motor Connected to a Six-Point Drum Switch.

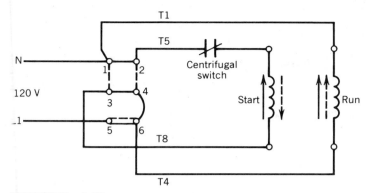

FIGURE 4-33

Split-Phase Motor Connected to a Six-Point Drum Switch.

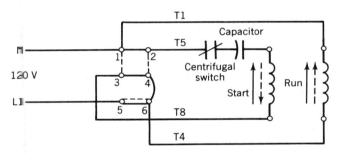

FIGURE 4-34

Single Voltage Capacitor-Start Motor Connected to a Six-Point Drum Switch.

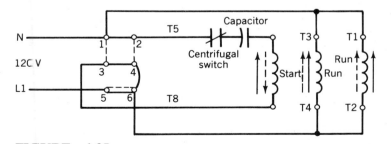

FIGURE 4-35

Dual Voltage Capacitor-Start Motor Connected to a Six-Point Drum Switch, 120 V Supply.

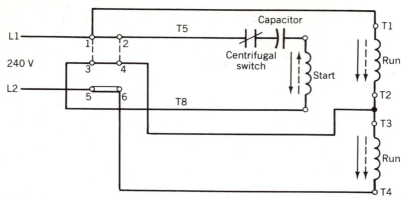

FIGURE 4-36

Dual Voltage Capacitor-Start Motor Connected to a Six-Point Drum Switch, 240 V Supply (method one).

It should be noted that when the drum switches for Figures 4-36 and 4-37 are in the off position, one live conductor is still connected to the motor windings. The electrical code in most areas will permit this, provided the control (drum switch) does not also serve as the disconnecting means.

Eight-Point Drum Switch

Figures 4-38 through 4-43 show the circuit diagrams for various AC motors controlled by eight-point drum switches. Ar-

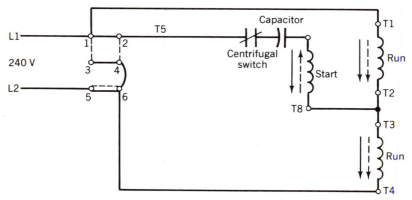

FIGURE 4-37

Dual Voltage Capacitor-Start Motor Connected to a Six-Point Drum Switch, 240 V Supply (method two).

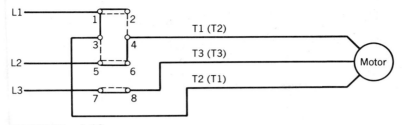

FIGURE   4-38
Three-Phase Motor Connected to an Eight-Point Drum Switch.

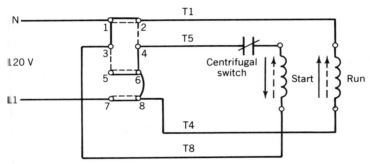

FIGURE   4-39
Split-Phase Motor Connected to an Eight-Point Drum Switch.

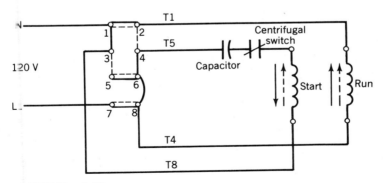

FIGURE   4-40
Single Voltage Capacitor-Start Motor Connected to an Eight-Point
Drum Switch.

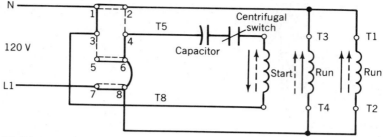

FIGURE 4-41

Dual Voltage Capacitor-Start Motor Connected to an Eight-Point Drum Switch, 120 V Supply.

rows have been used to indicate the direction of the instantaneous current flow.

Nine-Point Drum Switch

Figures 4-44 through 4-49 show the circuit diagrams for various AC motors controlled by nine-point drum switches. Arrows have been used to indicate the direction of the instantaneous current flow.

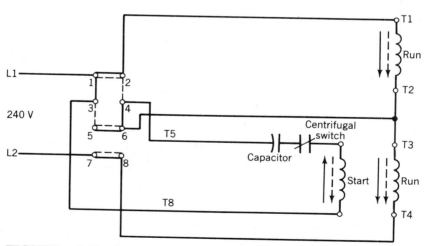

FIGURE 4-42

Dual Voltage Capacitor-Start Motor Connected to an Eight-Point Drum Switch, 240 V Supply (method one).

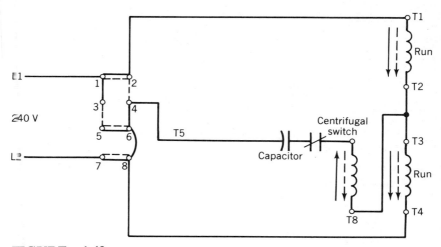

FIGURE 4-43

Dual Voltage Capacitor-Start Motor Connected to an Eight-Point Drum Switch, 240 V Supply (method two).

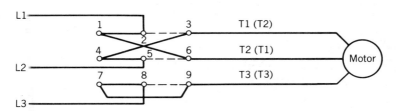

FIGURE 4-44

Three-Phase Motor Connected to a Nine-Point Drum Switch.

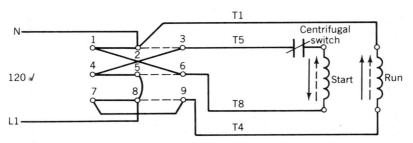

FIGURE 4-45

Split-Phase Motor Connected to a Nine-Point Drum Switch.

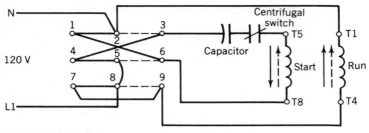

FIGURE 4-46

Single Voltage Capacitor-Start Motor Connected to a Nine-Point Drum Switch.

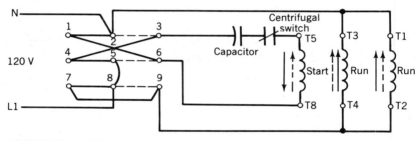

FIGURE 4-47

Dual Voltage Capacitor-Start Motor Connected to a Nine-Point Drum Switch, 120 V Supply.

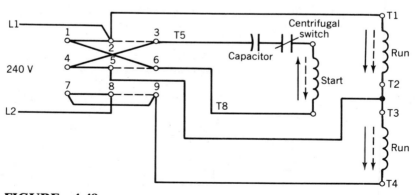

FIGURE 4-48

Dual Voltage Capacitor-Start Motor Connected to a Nine-Point Drum Switch, 240 V Supply (method one).

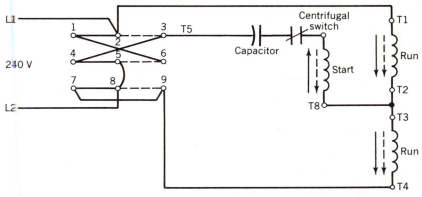

FIGURE 4-49

Dual Voltage Capacitor-Start Motor Connected to a Nine-Point Drum Switch, 240 V Supply (method two).

It should be noted that when the drum switches for Figures 4-48 and 4-49 are in the off position, one live conductor is still connected to the motor windings. The electrical code in most areas will permit this, provided the control (drum switch) does not also serve as the disconnecting means.

REVIEW EXERCISES

4-1 Explain why the secondary winding of a control transformer should be properly grounded.

4-2 Describe the construction of a dual-voltage relay.

4-3 Give an example in which a dual-voltage relay should be used.

4-4 Explain the sequence starting of motors.

4-5 Define "plugging."

4-6 Explain "anti-plugging."

4-7 Explain the function of a zero-speed switch.

4-8 Describe the construction of a zero-speed switch.

4-9 Compare the contact positions for the six-, eight-, and nine-point drum switches.

4-10 Explain the purpose for interlocking electrical equipment.

4-11 What is a ring circuit?

4-12 Explain how a zero-speed switch is physically installed on the equipment.

PROBLEM 4A
INTERLOCKING CIRCUIT

Design the control circuit based on the following instructions:

A. Use the physical layout shown in Figure 4-12.

B. Press the start button; all motor starter coils are energized at the same time.

C. Press the stop button; all motor starter coils are de-energized at the same time.

D. If an overload contact should open, only the motor in trouble will stop.

The solution to Problem 4A will be shown in the Instructor's Manual.

PROBLEM 4B
INTERLOCKING CIRCUIT

Design the control circuit for the installation shown in Figure 4-50. The circuit must operate as follows:

A. When thermostat no. 1 closes, motor no. 1 starts.

B. Following a 15-second delay, motor no. 2 starts (provided that thermostat no. 2 is in the closed position).

The solution to Problem 4B will be shown in the Instructor's Manual.

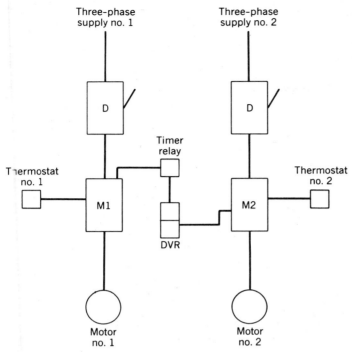

FIGURE 4-50

SECTION 5

TIMERS AND TRANSITION

INTRODUCTION TO SCHEMATICS

Sections 3 and 4 have been devoted to designing and reading schematics for full voltage starting. Understanding and being able to apply schematics is the backbone of motor control.

Troubleshooting must always commence with the schematic. Basic circuits may be serviced without referring to the schematic, but the service technician must understand the operation of the equipment and will have a mental picture of the schematic.

Sections 6 through 10 will cover reduced voltage and reduced current motor starting. Section 5 will cover topics that are common to the material discussed in those sections.

PURPOSE OF SCHEMATICS

The schematic is the method used when designing an electrical control circuit to convey to the electrician or technician how the equipment is to function.

The electrician or technician must be able to interpret the schematic in order to troubleshoot a piece of electrical equipment in an orderly and professional manner. The electrician must also be capable of translating the schematic diagram into a wiring diagram in order to install the equipment.

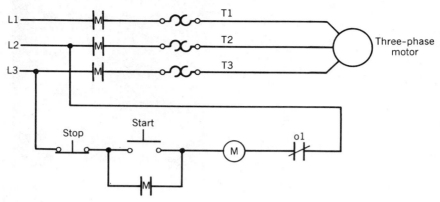

FIGURE 5-1

Schematic Diagram Power and Control Circuits for a Three-Phase Motor Installation—NVP Circuit.

A wiring diagram will enable the electrician to prepare a physical layout for the installation. Figures 5-1, 5-2, and 5-3 will demonstrate at the basic level the three steps just mentioned.

Figure 5-3 represents the physical layout of a typical motor installation. The disconnecting means, motor starter, control station, and the motor are all shown in their respective loca-

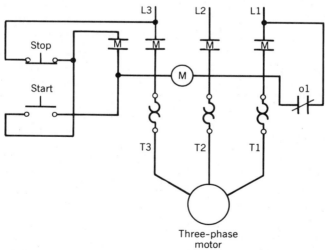

FIGURE 5-2

Wiring Diagram Power and Control Circuit for a Three-Phase Motor Installation—NVP Circuit.

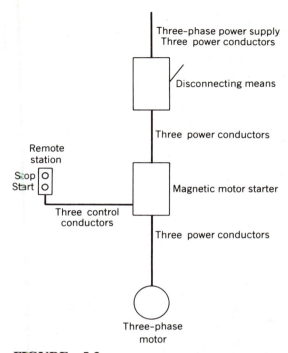

Three-phase power supply
Three power conductors

Disconnecting means

Three power conductors

Remote station

Stop
Start

Magnetic motor starter

Three control conductors

Three power conductors

Three-phase motor

FIGURE   5-3

Physical Layout for a Three-Phase Motor Installation—NVP Circuit.

tions. The physical layout will enable the electrician or technician to easily determine the requirements for the motor control installation. Take the time to make such a physical layout.

REDUCED VOLTAGE VERSUS REDUCED CURRENT MOTOR STARTING

The purpose of installing reduced voltage or reduced current starting is to restrict or limit starting current. When a motor is started on full line voltage, the starting current can be six times the nameplate current rating of the motor.

This abnormally high starting current could cause disturbances on the distribution lines. The power supply authority requires that the starting current be maintained at a tolerable level. These individuals should be consulted as to their requirements.

If resistors or autotransformers are used to reduce the starting voltage to the stator windings of a squirrel-cage motor, then the starting method is referred to as "reduced voltage starting." If resistors are inserted in the rotor circuit of a wound rotor motor during starting, the starting method is termed "reduced current starting." If the stator windings of a squirrel-cage motor are partly connected for starting (part-winding), or connected in one configuration for starting and another for running (wye-delta), the starting methods are also called "reduced current starting."

It is common practice to refer to all reduced voltage and reduced current starting methods as "reduced voltage starting."

In effect, each of the starting methods described will produce a reduced starting current, as opposed to full voltage starting current.

UNDERSTANDING SCHEMATICS

An excellent method of learning to read and understand schematic diagrams is to study the various power and control circuits for reduced voltage and reduced current starting.

The succeeding sections will deal with the various starting methods. Each section will deal with a specific method and related considerations, including

A.   The power circuit.
B.   The control circuit operation.
C.   Overcurrent protection.
D.   Overload protection.
E.   Selection of conductors or cables.
F.   Sizing of conduits.
G.   Coil sequence.
H.   Time delays.
I.   Transition.
J.   Troubleshooting techniques.

PNEUMATIC TIMERS

Pneumatic timers are widely used in motor control circuitry. Contacts are available with on-delay or off-delay. Contacts

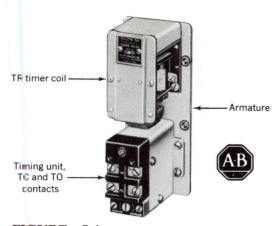

FIGURE   5-4
Pneumatic Timer.

are also available for timed to open (TO) or timed to close (TC). The time delay is adjustable, generally up to 180 seconds.

The timer relay illustrated in Figure 5-4 consists of a coil TR and a timing unit. When the TR coil is energized, the armature of the relay releases the pneumatic timing unit. After the preset time delay, the TO contact will open, and the TC contact will close.

The pneumatic timing unit illustrated in Figure 5-5 is much

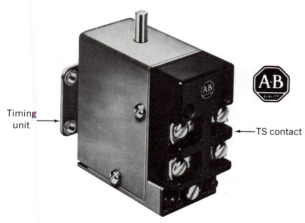

FIGURE   5-5
Pneumatic Timing Unit.

the same as a timer relay, the only difference being the absence of the TR coil. Placing the pneumatic timing unit under a contactor, which is located in a motor starter and used for another purpose, causes it to be triggered (released) when the contactor coil is energized. The triggered timed contact is generally called the TS (triggered switch) and must be identified as to which contactor coil triggered the TS contact.

COIL SEQUENCE

Under the heading "Designing a Control Circuit" in Section 3, the comment was made that coils are energized and deenergized. In reference to the two words "energized" and "deenergized," when a pilot device or any control device closes, a circuit is completed for a coil. This places the coil *into* the circuit. If the reverse happens and a pilot device opens, the circuit to the coil is broken, which takes the coil *out* of the circuit.

For simplicity, let us say that when a coil in a control circuit to be analyzed becomes energized, it is *in,* and when a coil is deenergized, it will be *out*. Note the two examples in Figures 5-6 and 5-7.

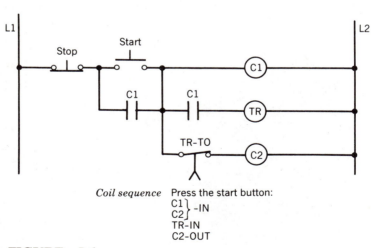

Coil sequence Press the start button:
C1 ⎫
C2 ⎬ –IN
TR-IN
C2-OUT

FIGURE 5-6

Coil Sequence (example one).

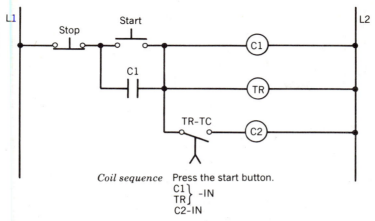

Coil sequence Press the start button.

C1 ⎫
TR ⎬ -IN
C2-IN

FIGURE 5-7

Coil Sequence (example two).

TRANSITION

The word transition means change. When discussing reduced voltage (current) motor starting, we see that there has to be a change from one state to another; for example, from reduced starting voltage to line running voltage, or from a starting configuration to a running configuration. The move from one mode to another is therefore termed a "transition." Two types of transition exist: open and closed.

Open transition means the motor is removed from the line during the changeover period; closed transition means the motor is not removed from the line during the changeover period.

Figure 5-8 is a theoretical graph to demonstrate how closed transition starting current could be plotted. The broken line indicates the possible current if the motor were started on full voltage. The solid line shows the possible current if the motor is started on reduced voltage (current). The solid line is not broken, indicating that the motor is never removed from the line during the changeover.

Figure 5-9 is a theoretical graph to demonstrate how open transition starting current could be plotted. The broken line indicates the possible current if the motor were started on full voltage. The solid line shows the possible current if the motor is started on reduced voltage (current). The solid line is bro-

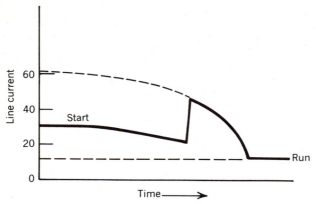

FIGURE 5-8

Graph Showing Starting and Running Current for Closed Transition.

ken, indicating that the motor was disconnected from the line during the changeover.

The information in Table 5-1 has been compiled from the various sections of this text, showing, at a glance, the types of transition for the various starting methods, and how to select the overload device for AC motors. For convenience, the applicable section of the text has also been listed.

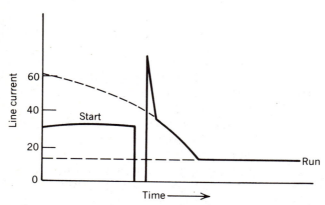

FIGURE 5-9

Graph Showing Starting and Running Current for Open Transition.

TABLE 5-1

Transition—Overload Selection Chart

Section	Starting Method	Transition	How to Select an Overload Device[a]
3	Full voltage	Does not apply	Apply FLA of the motor to the overload chart in motor starter.
8	Primary resis-tance	Closed	Apply FLA of the motor to the overload chart in motor starter.
10	Secondary resis-tance	Closed	Apply FLA of the motor to the overload chart in motor starter.
9	Wye-delta	Open or closed	Apply phase current to the overload chart in motor starter. (The phase current is motor FLA divided by $\sqrt{3}$.)
6	Autotransformer	*Manual,* open *Automatic,* open or closed	Apply FLA of the motor to the overload chart in motor starter.
7	Part-winding	Closed	Apply motor FLA divided by the number of parts to the overload chart in motor starter.

[a] Locate the overload device under the correct service factor (SF) column on the overload chart.

REVIEW EXERCISES

5-1 What is meant by the term "troubleshooting"?

5-2 Define "reduced current starting."

5-3 Define "reduced voltage starting."

5-4 Explain the operation of a pneumatic timer.

5-5 Explain the purpose for using a pneumatic timer.

5-6 How is the pneumatic timing unit installed?

5-7 Explain "coil sequence."

5-8 Explain "open transition."

5-9 Explain "closed transition."

5-10 Study the transition-overload selection chart. Use the chart when studying the various starting methods.

SECTION 6

AUTOTRANSFORMER STARTING

INTRODUCTION TO AUTOTRANSFORMER STARTING

The autotransformer method of starting provides reduced voltage to the stator windings at start, which will limit the starting current. Autotransformer starters are available with open or closed transition.

Autotransformer starting is so called because autotransformers are used in the power circuit to reduce the starting voltage. With the starting voltage reduced, the starting current will be lower than it would be if the motor were started on full line voltage. After a preset time delay, the autotransformers are removed from the circuit, and the squirrel-cage motor continues to run on line voltage.

As shown in Figure 6-1, an autotransformer consists of a single winding that is wound on a laminated iron core. Several taps may be extended from various points on the winding. The taps are identified as percentages.

The percentages will be of line voltage. The tap on the autotransformer shown in Figure 6-1 will be a percentage of line voltage. For example, if the supply voltage is 600 V and the percentage tap is 65%, then the voltage between the tap and T2 will be 390 V.

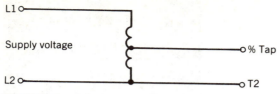

FIGURE 6-1

A Single Autotransformer.

Autotransformer starters may have two or three autotransformers for reducing the starting voltage. If two autotransformers are used, they will be connected, open delta, while three autotransformers will be connected wye (star).

Figure 6-2 illustrates the starting connection for an autotransformer starter using two autotransformers. Figure 6-3 depicts the starting connection for an autotransformer starter that uses three autotransformers.

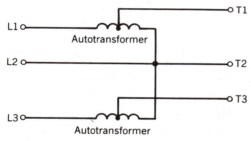

FIGURE 6-2

Two Autotransformers Connected Open Delta.

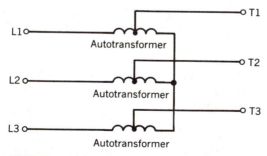

FIGURE 6-3

Three Autotransformers Connected Wye.

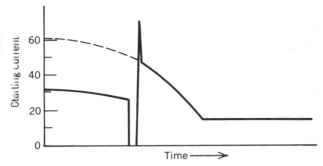

FIGURE 6-4

Graph Showing Starting and Running Current for Open Transition.

TRANSITION

Autotransformer starters may provide "open-circuit transition" or "closed-circuit transition" (see Section 5). Closed-circuit transition is preferred. The graphs in Figures 6-4 and 6-5 show what effect closed-circuit transition and open-circuit transition have on starting current.

AUTOTRANSFORMER
STARTER POWER CIRCUITS

Each motor control manufacturer will design power and control circuits that are different from all others. The tradesperson must learn to read and understand the various circuits, not necessarily attempt to memorize them.

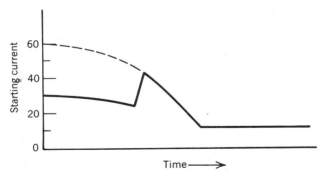

FIGURE 6-5

Graph Showing Starting and Running Current for Closed Transition.

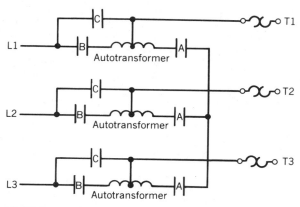

FIGURE 6-6

**Autotransformer Motor Starter
Power Circuit.**

Figures 6-6, 6-7, 6-8, and 6-9 are examples of just a few power circuits used by major motor control manufacturers.

With little effort, the tradesperson will learn to recognize Figures 6-6, 6-7, and 6-8 as closed-circuit transition, and Figure 6-9 as open-circuit transition. The four power circuits would be designed for magnetic autotransformer starters.

In addition, there are several power circuits for manual autotransformer starters. With all of the various power cir-

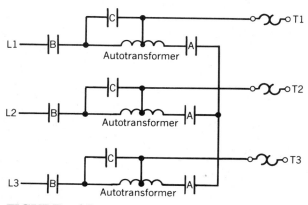

FIGURE 6-7

**Autotransformer Motor Starter
Power Circuit.**

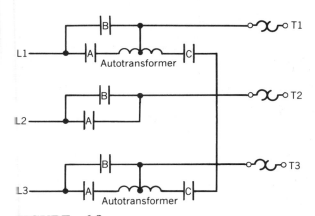

FIGURE 6-8

**Autotransformer Motor Starter
Power Circuit.**

cuits set aside, the starting connections will be either as
shown in Figure 6-2 or Figure 6-3.

PERCENTAGE TAPS

Autotransformers are supplied with percentage taps (mentioned earlier in this section). Generally, 50%, 65%, and 80%
taps are available. The starting current and starting *torque*

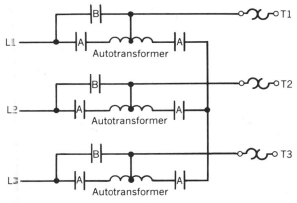

FIGURE 6-9

**Autotransformer Motor Starter
Power Circuit.**

may be controlled, depending on which tap is used. For example, the 50% voltage tap should provide the lowest starting current, but the motor may not have enough torque to rotate the driven load. If the 80% voltage tap were used, the starting torque may be adequate, but the starting current too high. Selecting the 65% voltage tap could limit the starting current to an acceptable level and still provide the required starting torque.

Which tap should you use? The tap that provides the best starting torque at the lowest starting current should be used. (*Note:* Insulate the unused taps.)

Also remember this: If a motor will not start, when started on line voltage, it will not start on reduced voltage.

OPEN TRANSITION
VERSUS CLOSED TRANSITION

The following comments are vital to the understanding of autotransformer starting. The circuits to be studied and analyzed are for magnetic autotransformer starters. The terms open- and closed-circuit transition will be discussed in detail. At this point, disregard all control circuits and focus only on the power circuits shown in Figures 6-10 and 6-11.

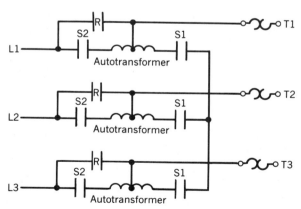

FIGURE 6-10

Autotransformer Motor Starter Power Circuit—Closed Transition.

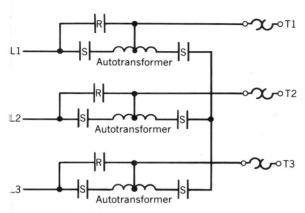

FIGURE 6-11

Autotransformer Motor Starter Power Circuit—Open Transition.

Open-Circuit Transition   This term means that the motor *is* removed from the line during the start-up period.

Closed-Circuit Transition   This term means that the motor *is not* removed from the line during the start-up period.

Circuit Operation, Figure 6-10

Contacts S1 and S2 will be closed for the first step and the autotransformers are connected in the wye configuration, as shown in Figure 6-3. Following the preset time delay, the S1 contacts will open, and for a split second, the circuit is completed from the lines through the S2 contacts, through the autotransformer windings to the T terminals to the motor winding. The R contacts will close, connecting the motor to the line and bypassing the autotransformers. Finally, the S2 contacts will open. The motor remains connected to the line throughout the starting sequence, therefore producing closed-circuit transition.

Circuit Operation, Figure 6-11

The S contacts will be closed for the first step and the auto-transformers are connected in the wye configuration, as shown in Figure 6-3. Following the preset time delay, the S contacts will open, disconnecting the motor from the line for a split second. The R contacts will close, thus connecting the motor to the line. When the S contacts open, open-circuit transition is produced.

ANALYZING CONTROL CIRCUITS

If we press the start button shown in Figure 6-12, a circuit is completed for the 1S coil. The N/O 1S contact closes and a

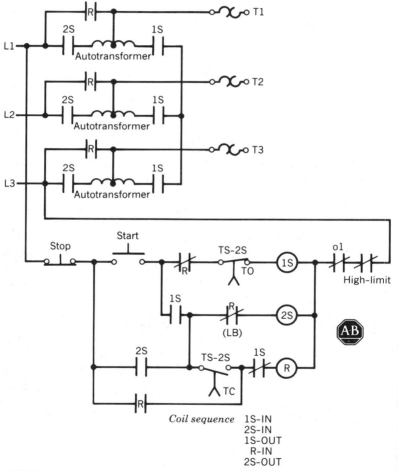

FIGURE 6-12

Autotransformer Motor Starter Power and Control Circuits—Closed Transition (example one).

circuit is completed for the 2S coil. The N/O 2S contact closes; it, along with the N/O 1S contact, maintains the 1S and 2S coils.

After a preset time delay, the contact TS–2S–TO opens and the TS-2S-TC contact closes. Coil 1S is deenergized, allowing the N/C 1S contact to reclose, completing the circuit for the R coil. The N/C R contact will open, deenergizing the 2S coil. At the same instant, the N/O R contact closes; it maintains the circuit for the R coil.

As for the power circuit, when the 1S and 2S coils are energized, the 1S and 2S power contacts close, starting the motor on reduced voltage through the autotransformers. When the 1S coil is deenergized, the 1S power contacts open and for a split second, the autotransformers are connected in series with the motor windings. The R power contacts close, connecting the motor to line voltage. The 2S power contacts are no longer required so they open when the 2S coil is deenergized. This is closed-circuit transition.

Circuit Operation, Figure 6-13

If we press the start button, a circuit is completed for the TR, 2S, and 1S coils at the same instant. The N/O 2S contact closes; it will maintain the circuit. The N/C 1S contact is now open. After a preset time delay, the TR–TO contact opens, deenergizing the 1S coil, allowing the N/C 1S contact to close. The TR–TC contact has closed and a circuit is completed for the R coil. The power circuit operation will be the same as illustrated in Figure 6-12.

Circuit Operation, Figure 6-14

If we press the start button, a circuit is completed for the S coil, closing the N/O S contact, energizing the TR coil. At the same instant, the N/C S contact opens. The N/O TR contact maintains the circuit. After a preset time delay, the TR–TO contact opens and the TR–TC contact closes. This action deenergizes the S coil, allowing the N/C S contact to close, completing a circuit for the R coil.

As for the power circuit, when the S coil is energized, all S power circuit contacts are closed. At this stage, the motor is connected to reduced voltage through the autotransformers. When the S coil is deenergized, the S power circuit contacts are open. The motor is taken off the line. Line voltage is applied to the motor terminals when the R coil is energized,

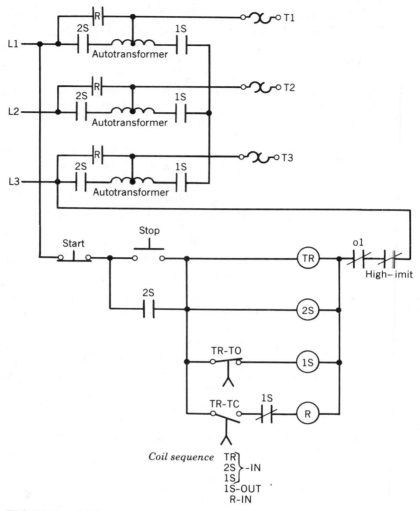

FIGURE 6-13

Autotransformer Motor Starter Power and Control Circuits—Closed Transition (example two).

closing the R power circuit contacts. This circuit is open transition.

HIGH-LIMIT AUTOTRANSFORMER PROTECTION

The autotransformers used in motor starting are for intermittent duty. A fault in the operation of the control circuit could

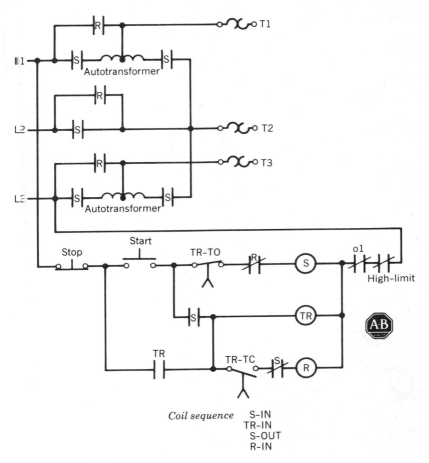

FIGURE 6-14

Autotransformer Motor Starter Power and Control Circuits—Open Transition.

cause the autotransformers to remain in the circuit. If this should occur, the autotransformers will heat up. In each of the previous circuits, high-limit, manual reset heat sensors have been installed in the control circuit to protect the autotransformers from becoming damaged.

MANUAL AUTOTRANSFORMER MOTOR STARTERS

When a squirrel-cage motor is not required to be automatically controlled, a manual autotransformer starter may be installed. The rugged construction readily lends itself to heavy industrial applications. The motor is started by moving a two-position handle, which is generally mounted on the right side of the starter. The handle is placed in the start position, and following a predetermined time delay, it is moved quickly to the run position. If this changeover is made slowly, the handle will not engage the run position. A mechanical device will prevent the handle from moving from start to run.

The motor is started on reduced voltage, through the auto-transformers. Line voltage is applied to the motor windings when the handle is in the run position. For a split second, as the contacts pass the center point, the motor is off the line. This will produce open transition.

Manual autotransformer starters may be used to start very large motors. Your motor control distributor will assist you in selecting the correct size for the installation.

Some manufacturers immerse the power contacts in oil to quench the arc, which occurs when the contacts make and break the circuit. Others use insulated barriers to prevent

FIGURE 6-15

Manual Autotransformer Motor Starter.

flashover. Figure 6-15 is a photo of a manual autotransformer starter that does not use oil to prevent flashover.

Operation of the Manual Autotransformer Starter, Figure 6-17

Placing the handle in the starting position will connect the autotransformers in open delta as shown in Figure 6-16. The motor is now starting on reduced voltage, which will reduce the starting current to the motor. Following a predetermined time delay, a bell will sound and the handle is quickly moved to the run position.

With this manual autotransformer starter, a latch coil is energized that will hold the handle in the running position. The starter provides no voltage protection.

It should be noted that the latch coil is not energized during the starting period. Some manufacturers energize the latch or holding coil during the starting period as well as the running position.

If an overload should occur, the overload contact shown in the control circuit of Figure 6-18 will open, deenergizing the latch coil C, releasing the spring-loaded handle to the off position.

The overload devices are not in the circuit during the starting period. This is acceptable for the following reasons:

1. Overload devices are intended to protect the motor during the run operation, not the starting period.

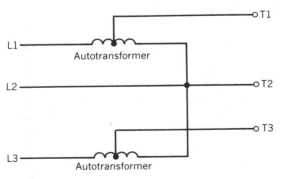

FIGURE 6-16

Two Autotransformers Connected Open Delta.

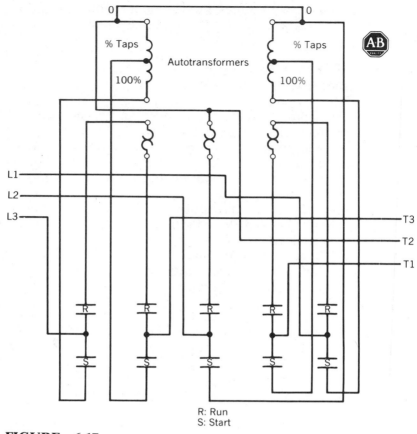

FIGURE 6-17

Manual Autotransformer Motor Starter Power Circuit Wiring Diagram.

2.   The electrical code will permit such a circuit provided
   A.   The starting handle will not remain in the starting position.
   B.   And the overcurrent device is in the circuit during the starting period.

No Voltage Protection?
No Voltage Release?
Which One?

Some confusion exists as to which category the manual autotransformer starter belongs. Textbooks refer to the

FIGURE   6-18

Schematic—Manual Autotransformer Motor Starter Control Circuit.

starters as no voltage release. What this means is that in the event of a supply voltage failure, the handle would be released to the off position. The autotransformer manual starters must then belong to the category of no voltage protection. The motor will not restart automatically on the resumption of the supply voltage.

POWER CIRCUIT CALCULATION

Single Motor Installation

To design the power circuit for a single motor installation, illustrated in Figure 6-19, the following calculations must be understood:

1. Overcurrent protection.
2. Overload protection.
3. Conductor sizes.

Overcurrent Protection

The branch-circuit overcurrent device is to be rated at not more than the percentage of the FLA rating of the motor shown in Table 6-1. Check this statement with the prepared table in the electrical code book having jurisdiction in your particular area. The prepared table will indicate the nearest size fuse or circuit breaker available.

Overload Protection

When selecting the correct overload device, obtain the full-load current from the motor nameplate. Apply the FLA to the overload chart, which is located on the cover or door of the motor starter, and select the proper overload device by catalogue number. Keep in mind the service factor of the motor. If the service factor is 1.0, the maximum value of the overload

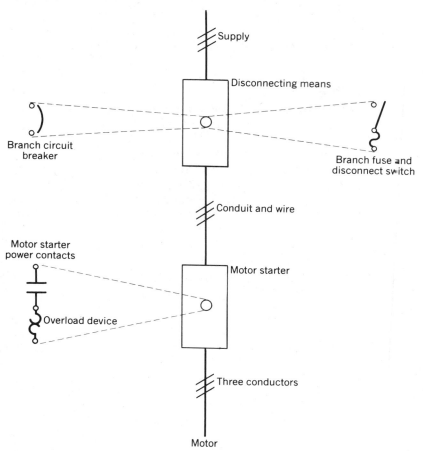

FIGURE 6-19

Single-Line Drawing of the Power Circuit for a Single Motor Installation Started by the Autotransformer Method.

device (in amperes) will be 115% of the FLA. With a service factor of 1.15, the maximum value of the overload device will be 125% of the FLA.

Conductor Sizes

Refer to Figure 6-12. It should be noted that with the auto-transformer method of starting, only three power circuit conductors are required.

Now look at Figure 6-19. The minimum allowable ampacity of the power circuit conductors will be 125% of the FLA of the motor.

TABLE 6-1
Canadian Electrical Code, Part 1

(TABLE 29)
(See Rules 28-200, 28-204, 28-208, and 28-210)*
RATING OR SETTING OF OVERCURRENT DEVICES FOR THE PROTECTION OF MOTOR BRANCH CIRCUITS

Type of Motor	Percent of Full-Load Current		
	Maximum Fuse Rating		Maximum Setting Time-Limit Type Circuit Breaker
	Time-Delay** "D" Fuses	Non-Time Delay	
Alternating Current			
Single-Phase, all types	175	300	250
Squirrel-Cage and Synchronous: Full-Voltage, Resistor and Reactor Starting	175	300	250
Auto-Transformer Starting: Not more than 30 A	175	250	200
More than 30 A	175	200	200
Wound Rotor	150	150	150
Direct Current	150	150	150

* (Except as permitted in Table 26 where 15-A overcurrent protection for motor branch-circuit conductors exceeds the values specified in the Table.)

** *Time delay "D" fuses are those referred to in Rule 14-200.*

NOTES: (1) *The ratings of fuses for the protection of motor branch circuits as given in Table 26 are based on fuse ratings appearing in the table above, which also specifies the maximum settings of circuit breakers for the protection of motor branch circuits.*

(2) *Synchronous motors of the low-torque low-speed type (usually 450 rpm, or lower) such as are used to drive reciprocating compressors, pumps, etc., and which start up unloaded, do not require a fuse rating or circuit-breaker setting in excess of 200% of full-load current.*

(3) *For the use of instantaneous trip (magnetic only) circuit interrupters in motor branch circuits see Rule 28-210.[a]*

[a] The (Table 29) and the Rule numbers mentioned refer to the Canadian Electrical Code. Similar comments will be listed in the National Electrical Code. With permission of Canadian Standards Association, this table is reproduced from CSA Standard C22.1-1982, Canadian Electrical Code, Part 1 (14th edition), which is copyrighted by CSA and copies may be purchased from CSA, 178 Rexdale Blvd., Rexdale, Ontario M9W 1R3.

If, for example, the nameplate FLA indicated 100 A, the minimum allowable ampacity of the conductors would be 100 × 125% = 125 A [for standardization, 90°C insulated conductor (copper) will be used in all examples]. The electrical code will require a minimum of no. 1 conductor.

REVIEW EXERCISES

6-1 What is the primary purpose for starting a squirrel-cage motor by the autotransformer method?

6-2 How are the autotransformers connected during the starting period when the autotransformer starter has three autotransformers?

6-3 How are the overload devices selected?

6-4 How many overload devices are required?

6-5 What is the purpose of the various voltage taps on the autotransformers?

6-6 Explain open and closed transition.

6-7 How is closed transition achieved with this method of starting?

6-8 What is the purpose of the high limit installed on the autotransformer?

6-9 Is this method of starting reduced voltage or reduced current starting?

6-10 How many power circuit conductors are required from the motor starter to the motor?

6-11 How many power circuit conductors are required from the disconnecting means to the motor starter?

6-12 Explain no voltage protection and give an example of the type of control circuit.

6-13 Explain no voltage release and give an example of the type of control circuit.

6-14 How is the minimum allowable ampacity of the power circuit conductors determined?

6-15 Explain what is meant by coil sequence.

PROBLEMS

The problems are divided into the following three categories:

Part A The application of the electrical code.
Part B Troubleshooting of motor control circuits.
Part C Multiple-choice questions on the combined topics.

Answers and solutions for the odd-numbered problems will be located at the back of this text. Answers and solutions for the even-numbered problems will be contained in the Instructor's Manual.

PART A
ELECTRICAL CODE

Instructions: When answering the following problems, keep in mind that

1. The starting method will be autotransformer starting.
2. The overload device will be the maximum value in amperes.
3. The conductors will be R90 X-Link CU.

Refer to Figure 6-19.

1. Calculate:
    A. The minimum conductor ampacity for the power circuit and

    B. Select the minimum size conductor for the power circuit when the FLA of the motor is 82 A.

2. If the FLA of a squirrel-cage motor is 52 A, calculate the

maximum value of the overload device when the service factor of the motor is 1.15.

_____

_____

3. Select the maximum allowable rating for a time-limit circuit breaker to be installed as the branch-circuit overcurrent device when the FLA of a squirrel-cage motor is 62 A.

_____

_____

4. Select the minimum size conduit required from the motor starter to the motor, for the wire size selected for problem no. 1.

_____

_____

PART B
TROUBLESHOOTING

Instructions: For this exercise, a fault will be stated. Study the designated circuit to determine how the circuit will operate.

1. See Figure 6-12. The N/C R contact in series with the 1S coil is defective and will not close. Press the start button. Discuss the results.

_____

_____

2. See Figure 6-12. The N/O R contact in the control circuit is broken and will not close. Press the start button. Discuss the results.

_____

_____

3. See Figure 6-12. While the motor is operating, an overload occurs. When the overload contact opens, how many coils would become deenergized?

_____

4. See Figure 6-12. The N/O 2S contact in the control circuit

will not close. Press the start button and hold it closed. Discuss the results.

_____

_____

_____

_____

5. See Figure 6-12. Repeat no. 4. Press the start button momentarily. Discuss the results.

_____

_____

_____

6. See Figure 6-13. The N/C 1S contact in the control circuit is defective and will not close. Press the start button. Discuss the results.

_____

_____

_____

_____

7. See Figure 6-13. The N/O 2S contact will not close. Press and hold in the start button. What will happen when the start button is released?

_____

_____

8. See Figure 6-14. The conductor on the left side of the TR coil is broken. Press the start button and then release it. Discuss the results.

_____

_____

PART C
MULTIPLE-CHOICE QUESTIONS

Instructions:

A. The electrical code book may be used to assist you in solving the following problems.

B. Select an answer from a, b, c, or d.

C. Answer all questions, based on the drawing in Figure 6-20.

1. The method of starting is:
   A. Wye-delta.
   B. Autotransformer.
   C. Full voltage.
   D. Secondary resistance.
2. The type of transition is:
   A. Overcurrent.
   B. Overload.
   C. Open.
   D. Closed.
3. The control circuit will provide:
   A. No field release.
   B. No voltage protection.
   C. Undervoltage release.
   D. No voltage release.
4. The overload device to protect the motor is selected by which method?
   A. FLA applied to the chart inside the motor starter.
   B. Cannot be properly selected.
   C. FLA × 300%.
   D. FLA divided by 2, applied to the code book.
5. When the starting sequence is completed, the motor is connected across line voltage. Which power contacts are closed?
   A. 2S and 1S.
   B. R.
   C. 2S and R.
   D. 1S and R.
6. The overload devices are installed in the power circuit to protect:
   A. The conductors during starting.
   B. The motor.
   C. The motor if a short-circuit occurs.
   D. The control circuit.

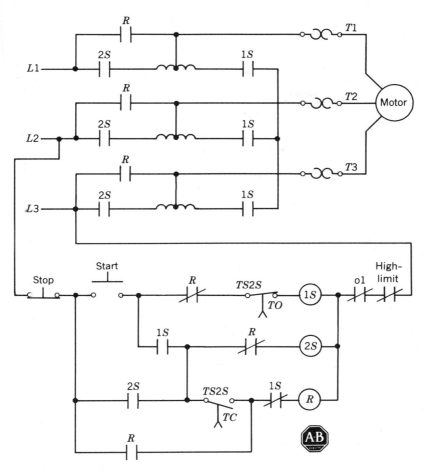

FIGURE 6-20

7. Press the start button, energizing the 1S coil. What is the next step in the operation of the control circuit?

 A. 1S N/O contact will close, and the 1S N/C contact will open.

 B. R N/C contacts will open.

 C. TS2S contact is activated.

 D. Nothing.

8. If the TS2S contact does not open after the preset time delay, what will happen to the 1S power contacts?

 A. Remain closed.

 B. Will open.

    C. Will vibrate.

    D. Will activate R contacts.

9. The power contacts that are mechanically interlocked are:

    A. 1S and R.

    B. 1S and 2S.

    C. 2S and R.

    D. 1S, 2S, and R.

10. If the branch-circuit overcurrent device is a time-delay fuse, the maximum percentage of FLA without special permission would be:

    A. 175%.

    B. 300%.

    C. 125%.

    D. 200%.

SECTION 7

PART-WINDING STARTING

INTRODUCTION TO PART-WINDING STARTING

Part-winding is a method of starting a squirrel-cage motor, which provides reduced current starting. Part-winding is so called because the motor is started by first connecting one part of the stator winding across line voltage. After a preset time delay, the second part of the stator winding is connected in parallel with the first part. By starting a two-part motor with this method, the starting current is limited to a lower level than it would be if both parts were connected across the line at the same moment.

POWER CIRCUITS

The power circuit shown in Figure 7-1 is two-stage (step). In operation, the S contacts must close first, with the R contacts closing after the preset time delay.

It is of the utmost importance to connect the motor terminals (T1, T2, T3, T7, T8, and T9) to the proper terminals on the motor starter. The motor winding (T1, T2, T3) must be

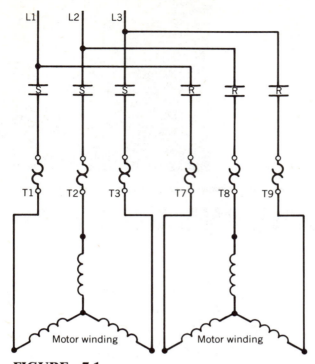

FIGURE 7-1

**Two-Part-Winding Power
Circuit.**

treated as a three-phase motor which when connected will have a definite rotation.

When the motor winding (T7, T8, T9) is connected, it will produce the same rotation. If by chance an error has been made, and T8 and T9 were interchanged, the second winding will attempt to change the rotation of the motor. Extremely high current will then flow, damaging the equipment.

Some motor control equipment manufacturers identify the load terminals of a part-winding motor starter using capital letters A through F. Care must be taken to check the motor terminal identification with the legend shown in their control manual.

Each motor control manufacturer will design a control circuit different from all others. With this in mind, it should be clear that a tradesperson must learn to read and understand the various circuits, and not attempt to just memorize the circuits.

Circuit Operation, Figures 7-1 and 7-2

First, look at Figure 7-2. If we press the start button, a circuit is completed for the S and TR coils. The N/O S contact closes, which maintains the circuit. After a preset time delay, the TR–TC contact closes, energizing the R coil.

As for the power circuit in Figure 7-1, when the S coil becomes energized, the S contacts will close. This connects the first part of the motor winding across line voltage. When the R coil becomes energized, the R contacts will close. This connects the second part across the line voltage, in parallel with the first part.

Throughout the start-up period, the motor never becomes disconnected from the supply. Therefore, the starting is closed transition.

Circuit Operation, Figures 7-1 and 7-3

First, look at Figure 7-3. If we press the start button, a circuit is completed for the S and TR coils. The N/O S contact closes and maintains the circuit. After a preset time delay, the TR–TC contact closes, energizing the R coil. The N/C R contact opens, deenergizing the TR coil. The N/O R contact is now closed, which maintains the circuit for the R coil. The power

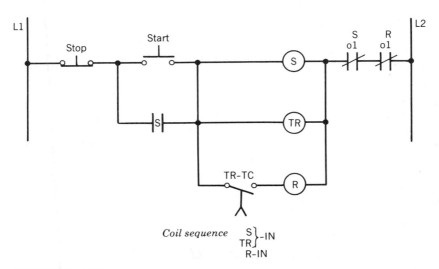

FIGURE 7-2
Two-Part-Winding Control Circuit (example one).

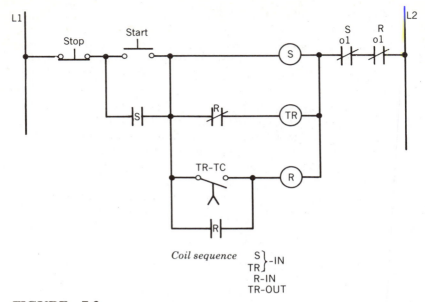

FIGURE 7-3

Two-Part-Winding Control Circuit (example two).

circuit in Figure 7-1 operates in the same manner as it did when it was controlled by the circuit in Figure 7-2.

Circuit Operation, Figure 7-4

Examine Figure 7-4. If we press the start button, a circuit is completed for the S coil. The N/O S contact closes, which maintains the circuit. After a preset time delay, the contact TS–S–TC (which is triggered by the S coil) closes, energizing the R coil.

POWER CIRCUIT CALCULATION

Single Motor Installation

Topics to be discussed:

1. Overcurrent protection.
2. Overload protection.
3. Conductor sizes.

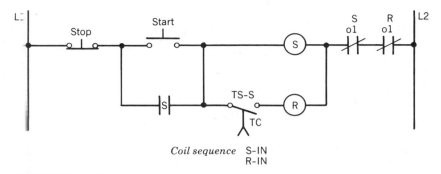

FIGURE 7-4
Two-Part-Winding Control Circuit (example three).

Overcurrent Protection

The branch circuit overcurrent device is to be rated at not more than 200% of the nameplate rating of the motor.

Overload Protection

The overload device must be selected to protect the winding it serves.

The current rating shown on the motor nameplate is for both windings; therefore, the FLA shown on the motor nameplate must be divided by the number of windings.

Example: If the FLA shown on the motor is 50 A, each winding will be 25 A. Apply 25 A to the chart inside the motor starter and select the correct overload device. Keep in mind the SF (service factor) of the motor.

Conductor Sizes

To select the conductors to the supply side of the motor starter, the minimum ampacity of the conductors must be 125% of the motor FLA. To select the conductors from the load side of the motor starter to the motor terminals, the minimum ampacity of the conductors must be 125% of the current rating of the winding it serves.

It should be noted that three conductors are required to the supply side of the motor starter, and six conductors are required from the motor starter to the motor terminals.

Various wiring methods may be used to complete the power circuit installation. The following will give you instruction for three possible installations.

POWER CIRCUIT INSTALLATION FOR VARIOUS WIRING METHODS

Single Motor Installations

Figure 7-5 illustrates an installation using single conduit runs and 90° copper conductors.

To calculate the minimum ampacity of the conductors to the supply side of the motor starter, the nameplate current rating of the motor must be used.

Example: If the motor nameplate indicated 100 A, the minimum allowable ampacity of the conductors must be 100 A × 125% = 125 A. The electrical code will require a minimum of no. 1 conductors.

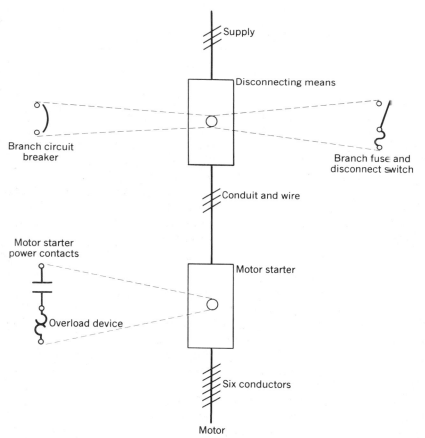

FIGURE 7-5

Single-Line Drawing of the Power Circuit for a Single Motor Installation Started by the Part-Winding Method.

To calculate the minimum allowable ampacity of the conductors from the motor starter to the motor, the conductors must have a minimum ampacity of 125% of the winding they serve.

Example: Each winding of the 100-A motor will be rated at 50 A. Therefore, the conductors must have an ampacity of 50 A × 125% = 62.5 A. The code will require six no. 4 conductors. The ampacity of a no. 4 conductor is 90 A, but with six conductors installed in the same raceway, the ampacity of the conductors will be reduced to 90 A × 80% = 72 A. The required ampacity is 62.5 A; thus, no. 4 is acceptable.

Figure 7-6 represents an installation using multiple conduit runs from the motor starter to the motor.

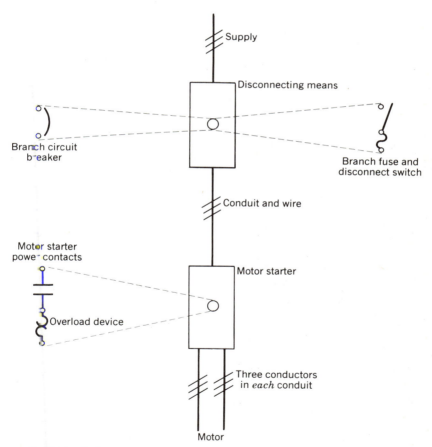

FIGURE 7-6

Single-Line Drawing of the Power Circuit for a Single Motor Installation Started by the Part-Winding Method.

If we use the same data as in the previous example (Figure 7-5), the conductors to the supply side of the motor starter will be the same size as in that diagram.

However, the conductors from the motor starter to the motor may be smaller. The ampacity of the conductors need not be derated, as there are only three conductors in each conduit. Therefore, six no. 6 conductors may be installed in each conduit. One conduit will contain conductors from T1, T2, and T3, while the second conduit will contain conductors from T7, T8, and T9.

Figure 7-7 depicts an installation using single conductor R90 Cu.

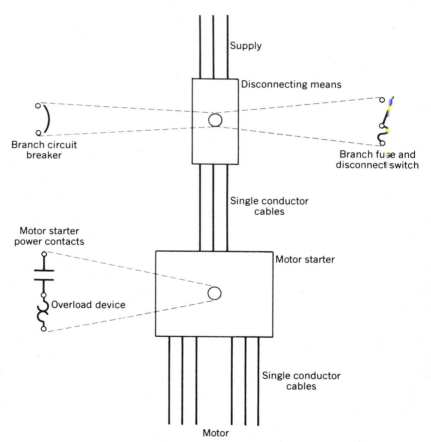

FIGURE 7-7

Single-Line Drawing of the Power Circuit for a Single Motor Installation Started by the Part-Winding Method.

To calculate the minimum ampacity of the conductors to the supply side of the motor starter, the nameplate current rating of the motor must be used.

Example: If the FLA were 200 A, the minimum allowable ampacity of the conductors would be 200 A × 125% = 250 A. No. 2/0 CU R90 would be required.

To calculate the minimum ampacity of the conductors from the motor starter to the motor, the current rating of each winding must be used. Each winding would be rated at 100 A. Therefore, 100 A × 125% = 125 A. No. 4 CU R90 conductors will be required.

Two-Part, Three-Stage Starting

Having completed the various circuits for two-part (two-stage) starting, let us discuss the two-part, three-stage starting.

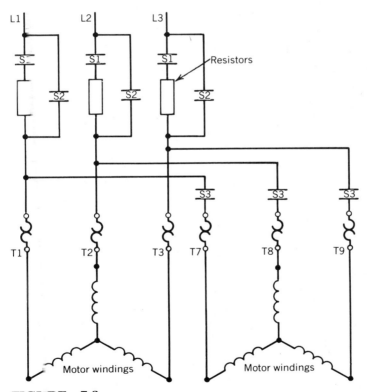

FIGURE 7-8

Three-Stage, Two-Part-Winding Power Circuit.

The motor is a two-part winding motor. The difference is the inclusion of a set of resistors in the power circuit.

Refer to the power circuit in Figure 7-8. In operation, the power circuit contacts S1 will close, which connects the resistance units in series with the first (part) winding (T1, T2, T3). This action will limit the starting current to a lower value than it would be if the first winding were connected across line voltage. After a preset time delay, the power contacts S2 will close, which removes the resistance units from the circuit. The first winding is now connected to line voltage. Following the second time delay, the S3 power contacts will close, connecting the second (part) winding in parallel with the first winding.

The control circuit shown in Figure 7-9 is an example of a circuit that could be designed to control the three-stage, two-part-winding power circuit illustrated in Figure 7-8.

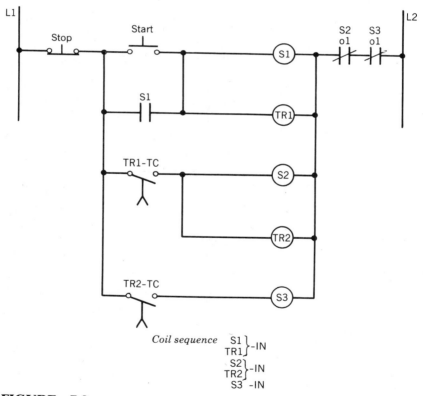

FIGURE 7-9

Three-Stage, Two-Part-Winding Control Circuit (example one).

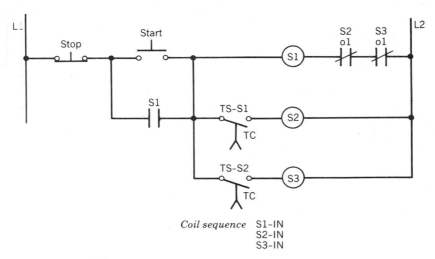

FIGURE 7-10

Three-Stage, Two-Part-Winding Control Circuit (example two).

If we press the start button, a circuit is completed for the S1 and TR1 coils. The S1 coil will close the S1 power contacts, plus the N/O S1 contact in the control circuit, which will maintain the circuit. After a preset time delay, the TR1–TC contact closes, which will energize the S2 and TR2 coils. The S2 power contacts will close. After a second time delay, the TR2–TC contact closes, energizing the S3 coil. The S3 power contacts will now be closed. The motor windings are now operating in parallel on line voltage.

Another circuit is presented in Figure 7-10 that could be used to operate the power circuit shown in Figure 7-8. If we press the start button, a circuit is completed for the S1 coil. The N/O S1 contact maintains the circuit. The TS–S1–TC contact will complete the circuit for the S2 coil; the TS–S2–TC contact will complete the circuit for the S3 coil.

REVIEW EXERCISES

7-1   What is the purpose of using the two-part-winding method of starting?

7-2   Explain the operation of the power circuit for the two-part-winding method of starting.

7-3   Explain the operation of the power circuit for the three-stage, two-part-winding method of starting.

7-4   Is part-winding starting open or closed transition?

7-5   How many power circuit conductors are required from the motor starter to the motor?

7-6   How many power circuit conductors are required to the supply side of the motor starter?

7-7   How are the overload devices selected for the two-part-winding method of starting?

7-8   How are the overload devices selected for the three-stage, two-part-winding method of starting?

7-9   How are the conductor sizes selected on the supply side of the motor starter?

7-10   How are the conductor sizes selected from the motor starter to the motor for the two-part-winding method of starting?

7-11   Is two-part-winding starting reduced voltage or reduced current starting?

7-12   Define overcurrent protection.

7-13   Define overload protection.

7-14   Explain the operation of a timing relay that incorporates a TR coil.

7-15   Explain the operation of a timing unit that is released or triggered by a coil other than a TR coil.

PROBLEMS

The problems are divided into the following three categories:

Part A   The application of the electrical code.
Part B   Troubleshooting of motor control circuits.
Part C   Multiple-choice questions on the combined topics.

Answers and solutions for the odd-numbered problems will be located at the back of this text. Answers and solutions for the even-numbered problems will be found in the Instructor's Manual.

PART A
ELECTRICAL CODE

Instructions: When answering the following problems, keep in mind:

1. The starting method will be two-part-winding.
2. The overload device will be the maximum value in amperes.
3. The conductors will be R90 X-Link CU.

Refer to Figure 7-5.

1. Calculate:
   A. The minimum conductor ampacity for the power circuit and

   _____

   _____

   B. Select the minimum size conductor for the power circuit from the motor starter to the motor when the FLA of the motor is 80 A.

   _____

2. The FLA of a squirrel-cage motor is 40 A. Calculate the maximum value of the overload device when the service factor of the motor is 1.15.

   _____

3. The FLA of a squirrel-cage motor is 33 A. Calculate the maximum value of the overload device when the service factor of the motor is 1.0.

   _____

4. Select the minimum size conductor for the power circuit to supply the line side of the motor starter. The motor FLA is 100 A.

   _____

   _____

PART B
TROUBLESHOOTING

Instructions: For this exercise, a fault will be stated. Study the designated circuit to determine how the circuit will operate.

1. See Figure 7-2. The conductor on the left side of the TR coil has been broken. Press the start button. Discuss the results.

    _____

    _____

    _____

2. See Figure 7-2. The TR–TC contact is welded shut. Press the start button. Discuss the results.

    _____

    _____

    _____

3. See Figure 7-3. The N/O S contact is defective and will not close. Press the start button. Discuss the results.

    _____

    _____

    _____

4. See Figure 7-3. The N/C R contact is defective and will not open. Press the start button. Discuss the results.

    _____

    _____

    _____

5. See Figure 7-3. The N/C R contact is defective and remains in the open position. Press the start button. Discuss the results.

    _____

    _____

    _____

6. See Figure 7-3. The N/C R contact and the N/O R contact are both defective. In operation, the contacts remain in

the same positions. Press the start button. Discuss the results.

_____

_____

_____

7. See Figure 7-4. The S coil has an open circuit. Press the start button. Discuss the results.

_____

_____

8. See Figure 7-4. The TS–S contact is defective and will not close. Press the start button. Discuss the results.

_____

_____

_____

PART C
MULTIPLE-CHOICE QUESTIONS

Instructions:

A. The electrical code book may be used to assist you in solving the following problems.

B. Select an answer from a, b, c, or d.

C. Answer all questions, based on the drawing in Figure 7-11.

1. The method of starting is:
   A. Wye-delta.
   B. Full voltage.
   C. Part-winding.
   D. Primary resistance.

2. The type of transition is:
   A. Closed.
   B. Open.

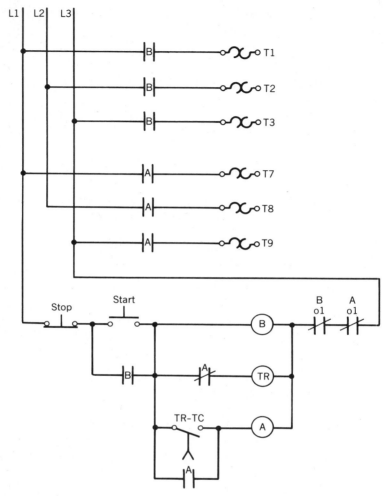

FIGURE 7-11

    C. Field protection.

    D. NVP

3. The control circuit provides:

    A. No voltage release.

    B. Undervoltage release.

    C. Low voltage release.

    D. No voltage protection.

4. The overload device is selected by which method?

    A. FLA × 250%.

    B. FLA divided by the number of parts, taken to the chart in motor starter.

    C. FLA to chart in motor starter.

    D. Cannot be selected without the wire size.

5. When the motor is running properly, how many contacts in the power circuit are open?

    A. 0.

    B. 6.

    C. 3.

    D. 2.

6. During the starting period, the instant before the TR–TC contact closes, how many contacts in the control circuit are open?

    A. 2.

    B. 3.

    C. 1.

    D. 4.

7. When the N/C A contact in the control circuit opens, what is the next sequence of operation?

    A. The motor picks up speed.

    B. The B coil seals in.

    C. The TR coil is deenergized.

    D. The A coil is deenergized.

8. If the overload contact for the B starter should open, which coil or coils will become deenergized at the same instant?

    A. TR.

    B. TR and B.

    C. A and B.

    D. TR and A.

SECTION 8

PRIMARY RESISTANCE STARTING

INTRODUCTION TO PRIMARY RESISTANCE STARTING

Primary resistance starting is so called because preengineered resistance units are connected in the line to the stator winding of a squirrel-cage motor during the start-up period. The primary circuit of a squirrel-cage motor is the stator winding. A voltage drop occurs across the resistors during the start-up period, which provides a reduced voltage across the stator windings. With the starting voltage reduced, the starting current is lower than it would be if the motor were started on full line voltage. After a preset time delay, the resistance units are shunted out of the circuit, and the motor continues to run on line voltage.

This method of starting is available for two-, three-, or four-stage starting.

TWO-STAGE PRIMARY RESISTANCE STARTING

In operation, the power circuit shown in Figure 8-1 would involve the following: The S contacts must close first, with the R contacts closing after the preset time delay.

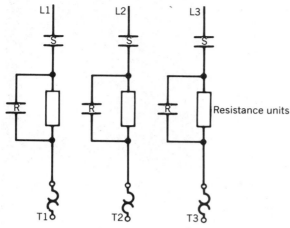

FIGURE 8-1

**Two-Stage Primary Resistance
Power Circuit (method one).**

Circuit Operation, Figure 8-2

The control circuit shown in Figure 8-2 will control the power circuit shown in Figure 8-1. When we press the start button, a circuit is completed for the S and TR coils. The N/O S contact closes, which maintains the circuit. After a preset time delay, the TR–TC contact closes, energizing the R coil.

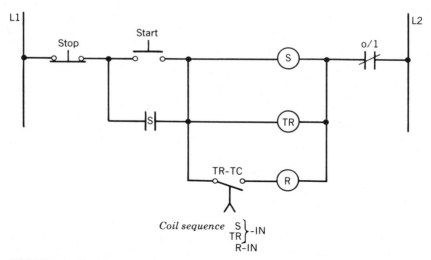

FIGURE 8-2

Two-Stage Primary Resistance Control Circuit (example one).

When the S coil is energized, the S power contacts close, connecting the resistance units in series with the stator winding. When the R coil is energized, the R power contacts close, shunting the resistors out of the circuit.

Throughout the starting period, the motor never becomes disconnected from the supply. Therefore, the starting method is closed transition.

Circuit Operation, Figures 8-3 and 8-1

The control circuit shown in Figure 8-3 will control the power circuit illustrated in Figure 8-1. Pressing the start button completes a circuit for the S and TR coils. The N/O S contact in the control circuit maintains the circuit. After a preset time delay, the TR–TC contact closes, energizing the R coil. The N/C R contact opens, deenergizing the TR coil. The N/O R contact maintains the circuit for the R coil.

The power circuit shown in Figure 8-1 operates in the same manner as it did when controlled by the circuit shown in Figure 8-2.

Circuit Operation, Figures 8-4 and 8-5

Look at Figure 8-5. If we press the start button, a circuit is completed for the S and TR coils. The N/O S contact closes,

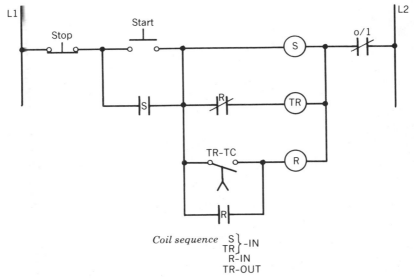

FIGURE 8-3

Two-Stage Primary Resistance Control Circuit (example two).

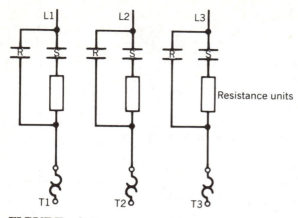

FIGURE 8-4

**Two-Stage Primary Resistance
Power Circuit (method two).**

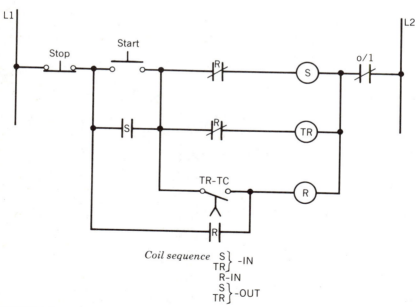

FIGURE 8-5

Two-Stage Primary Resistance Control Circuit (example three).

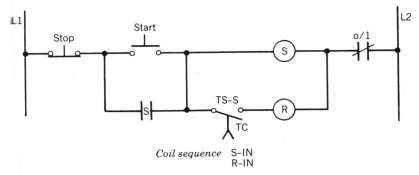

FIGURE 8-6

Two-Stage Primary Resistance Control Circuit (example four).

which maintains the circuit. After a preset time delay, the TR–TC contact closes, energizing the R coil. The N/C R contacts will open, deenergizing the S and TR coils. The N/O R contact closes, which maintains the circuit for the R coil.

As for the power circuit in Figure 8-4 when the S coil becomes energized, the S contacts in the power circuit will close. This action connects the resistance units in series with the stator windings. When the R coil becomes energized, the R contacts will close. This will connect the motor to line voltage. When the S coil is deenergized, the S contacts open, which removes the resistors from the circuit.

Circuit Operation, Figure 8-6

Look at Figure 8-6. If we press the start button, a circuit is completed for the S coil. After a preset time delay, the TS–S–TC contact closes, energizing the R coil.

THREE-STAGE PRIMARY RESISTANCE STARTING

Figure 8-7 illustrates the power circuit for a three-stage primary resistance starter. At the beginning of this section, it was mentioned that primary resistance starters have been and are still available for three-stage starting. Two sets of resistance units are connected in series with the stator (primary) winding of the motor on the first stage. After a preset time delay, the first set of resistance units are removed from the

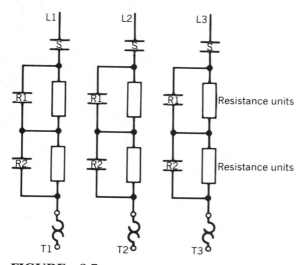

FIGURE 8-7

**Three-Stage Primary Resistance
Power Circuit (method one).**

circuit. Following the second time delay, the remaining set of resistors are removed from the circuit. The motor continues to run on line voltage.

The control circuit illustrated in Figure 8-8 will operate the power circuit in Figure 8-7.

Pressing the start button completes a circuit for the S and TR1 coils. The N/O S contact closes, which will maintain the circuit. After a preset time delay, the TR1–TC contact closes, energizing the R1 and TR2 coils. Following a second time delay, the TR2–TC contact closes, energizing the R2 coil.

It should be noted that when the S, R1, and R2 coils are energized, the respective contacts in the power circuit close. Additional contacts could be installed to remove the TR1 and TR2 coils, as well as contacts to maintain the S, R1, and R2 coils.

The circuit discussed in Figure 8-8 is one that could have been designed by a motor control equipment manufacturer. Thus, we would not be involved in the circuit design but rather the maintenance of the equipment. If the equipment should fail to function properly, by using the proper schematics, the fault can be easily located and repairs made.

The circuit in Figure 8-8 was designed using a pneumatic timer, controlled by a coil. Some control circuit designers will choose to use a pneumatic timing unit, which is located below

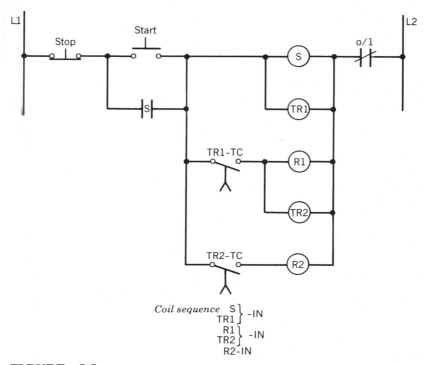

FIGURE 8-8

Three-Stage Primary Resistance Control Circuit (example one).

a coil, used for another purpose, in the motor starter. The timer contact is then referred to as a triggered contact (referred to as TS).

We as tradespeople should not be overly critical of any particular motor control company, as control circuits may be altered for the sake of change. This does not mean that the circuit is incorrect. We are entitled to an opinion, but it is more constructive to learn to read the schematics and to train ourselves to troubleshoot the circuits as installed.

The control circuit shown in Figure 8-9 will operate the power circuit in Figure 8-7.

The TR coils have been deleted and replaced by pneumatic timing units. The pneumatic timing units have been installed below and will be triggered by the respective coil armatures. The identification TR should now be dropped from the time-delay contacts. The identification will now be TS for triggered switch.

When we press the start button, a circuit is completed for

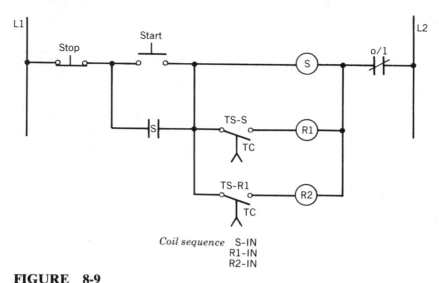

FIGURE   8-9

Three-Stage Primary Resistance Control Circuit (example two).

the S coil. The N/O S contact closes, which will maintain the circuit for the S coil. The TS–S–TC contact is triggered by the S coil. After a preset time delay, the TS–S–TC contact will close, energizing the R1 coil. The TS–R1–TC is triggered by the R1 coil. After a second time delay, the TS–R1–TC contact will close, energizing the R2 coil.

It should be noted that when the S, R1, and R2 coils are energized, the respective contacts in the power circuit close.

FOUR-STAGE PRIMARY RESISTANCE STARTING

Figure 8-10 depicts a power circuit for a four-stage primary resistance starter. This circuit is an extension of the three-stage starter shown in Figure 8-7. It should be noted that when the motor is running on line voltage, all power contacts are closed (S, R1, R2, and R3). This type of power circuit is used by some motor control circuit designers. Others prefer the power circuit illustrated in Figure 8-11, which only requires the power contacts R3 to be closed when the motor is running.

Figure 8-12 is a control circuit that could be used to operate the power circuits shown in Figures 8-10 or 8-11.

The designer of the control circuit in Figure 8-13 has chosen

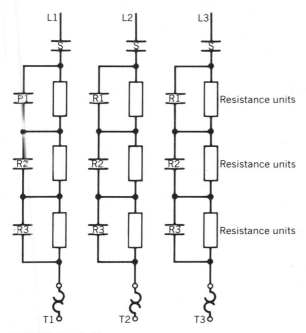

FIGURE 8-10

Four-Stage Primary Resistance Power Circuit (method one).

not to use TR coils to activate the time-delay contacts. Again, the time-delay contacts are triggered by a coil located within the starter, which is used for another purpose. This is similar to the circuits shown in Figures 8-6 and 8-9.

The control circuit in Figure 8-13 may be used to operate the power circuits shown in Figures 8-10 and 8-11.

Another example is presented in Figure 8-14. This control circuit could also be used to operate the power circuit shown in Figure 8-11.

When we press the start button, a circuit is completed for the S coil, closing the N/O S contact, which will maintain the circuit for the S coil. Following the first time delay, the TS–S–TC contact will close, completing the circuit for the R1 coil. The TS–R1–TC contact will close following the second time delay, completing the circuit for the R2 coil. The circuit for the R3 coil is completed when TS–R2–TC closes. The N/C R3 contact will open, deenergizing the S coil. The R1 and R2 coils are deenergized at the same instant. The N/O R3 contact maintains the circuit for the R3 coil.

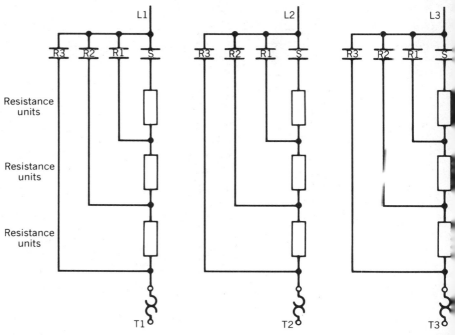

FIGURE 8-11

Four-Stage Primary Resistance Power Circuit (method two).

When the motor is running on line voltage, the only power contacts closed will be the R3 contacts.

ANALYZING A CONTROL CIRCUIT

The ideas expressed under this heading could be applied to any starting method or control circuit.

The circuits in Figures 8-15, 8-16, 8-17, and 8-18 will be used to demonstrate how circuits may be altered and to discuss the results of such changes.

Figure 8-15 illustrates the power circuit of a two-stage primary resistance starter. In operation, the S contacts will close, and after a preset time delay, the R contacts will close.

The circuit shown in Figure 8-16 is similar to the circuit used by a major motor control company. If the circuit is studied, it will be noticed that if for any reason the TR coil does not become energized, the motor will not start. However, it has been suggested that the TR contact (identified F) is redun-

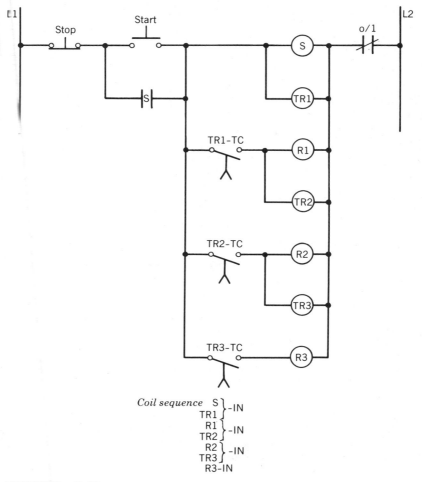

FIGURE 8-12

Four-Stage Primary Resistance Control Circuit (example one).

dant and is not required. Could it be removed and the circuit altered to resemble Figure 8-17?

Study the circuit in Figure 8-17. Note that the TR contact identified as B has been removed and a conductor A installed. Form an opinion. Is the new altered circuit acceptable? Before you make a decision, ask the following two questions:

Question 1   If the motor is running and an overload occurs, the overload contact will open. Will the motor stop? The answer must be *yes*.

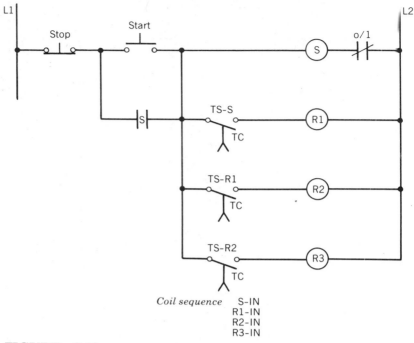

FIGURE 8-13

Four-Stage Primary Resistance Control Circuit (example two).

Question 2   If the overload contact is open and the motor is not running, press the start button. Will the motor start? The answer must be *no*.

When you apply the two questions to the control circuit in Figure 8-16, the answers will be yes and no. The control circuit is therefore acceptable. If you apply the two questions to the control circuit in Figure 8-17, the answers will be yes and yes. The control circuit is therefore unacceptable.

The control circuit in Figure 8-18 is the result of a further alteration to the original circuit. When we apply the two questions above to the new circuit, the answers will be yes and no. The control circuit is therefore acceptable.

Control circuits should not be altered just for the sake of making a change. If a circuit must be altered, do so very carefully, so as not to create a problem in another part of the power or control circuits.

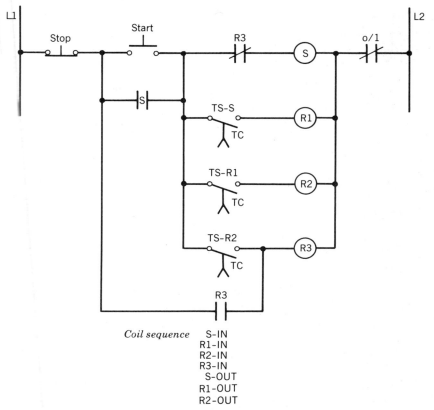

FIGURE 8-14

Four-Stage Primary Resistance Control Circuit (example three).

MANUAL PRIMARY RESISTANCE STARTING

In addition to automatic and semiautomatic primary resistance starting, manual primary resistance starters are available. Unlike the automatic or semiautomatic starters, the starting of the motor is entirely under the control of the operator with a manual starter.

Figure 8-19 is a photo of such a motor starter. Figure 8-20 shows a wiring diagram of the power and control circuits of a manual primary resistance stepless motor starter.

The manual primary resistance starter uses graphite com-

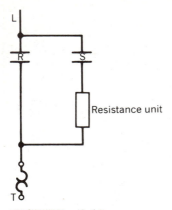

FIGURE 8-15

**Two-Stage Primary Resistance
Power Circuit.**

pression disc resistors placed in insulated steel tubes. Figure
8-21 represents a cutaway view of the resistor.

Figure 8-20 shows three resistance units, one in each line.
Each resistance unit consists of a column of graphite discs. At
the top of each tube is a pressure plug, electrode, and flexible
jumper, and the second terminal is attached to the bottom of
the tube.

Current flowing through the discs encounters little resistance from the graphite itself, but the treated surfaces of the

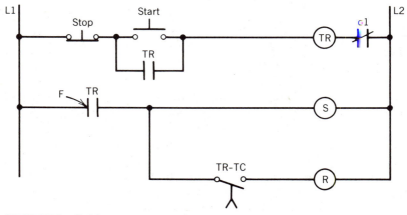

FIGURE 8-16

Two-Stage Primary Resistance Control Circuit.

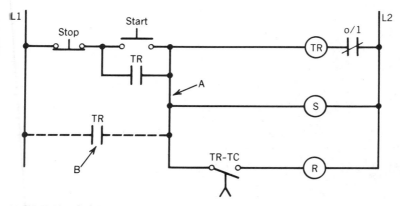

FIGURE 8-17

Two-Stage Primary Resistance Control Circuit (altered).

discs offer definite contact resistance. Increasing or decreasing the amount of pressure applied to the discs will decrease or increase, respectively, the total ohmic value of the resistor.

Starter Operation

Lifting the starting lever quickly to the mid-point ensures a completed power circuit for the motor through all three compression disc resistors.

The resistors are now at maximum resistance. The starting current and starting torque are quite low, which eliminates

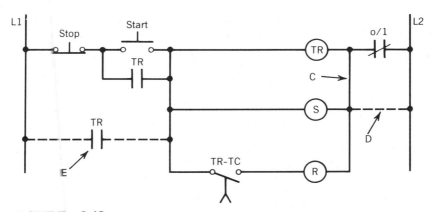

FIGURE 8-18

Two-Stage Primary Resistance Control Circuit (altered).

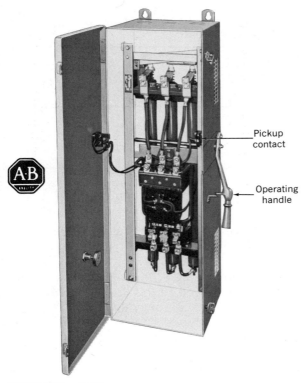

Pickup contact

Operating handle

FIGURE 8-19

Manual Primary Resistance Motor Starter.

line disturbances and excessive starting torque for the driven machinery. Lifting the starting level further increases the pressure on the discs, which decreases the resistance of the resistors, and increasing voltage at the motor terminals.

When the handle reaches the top of the travel, the resistors are bypassed, and the motor continues to run on line voltage.

The pickup contact energizes the M coil, which closes the M power contacts, which bypass the starting resistors. Figure 8-20 indicates that the pickup contact is N/O. This is not correct. The drawing shows the actual installed condition. In fact, the pickup contact is N/C, held open. When the starting lever reaches the top of the travel, it does not close the pickup contact, but rather allows it to close.

The starting handle is spring-loaded to the off position. If an overload should occur, the overload contact will open, de-

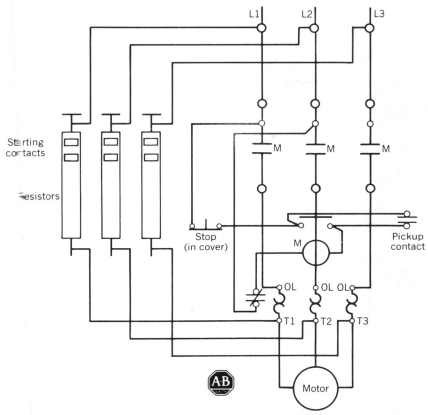

FIGURE 8-20

Manual Primary Resistance Power Circuit.

energizing the M coil, opening the M contacts, stopping the motor. Pressing the stop button will also deenergize the M coil, stopping the motor. The motor starter provides no voltage protection.

It should be noted that the overload devices are not in the circuit during the starting period. This is acceptable for the following reasons:

1. Overload devices are intended to protect a motor during the running operation, and not the starting period.

2. The electrical code permits this practice provided

   A. The starting handle will not remain in the starting position.

FIGURE   8-21
Compression Disc Resistor.

B.   And the overcurrent protection is in the circuit during the starting period.

POWER CIRCUIT CALCULATION

Single Motor Installation

To design the power circuit for a single motor installation as illustrated in Figure 8-22, the following calculations must be understood:

1.   Overcurrent protection.

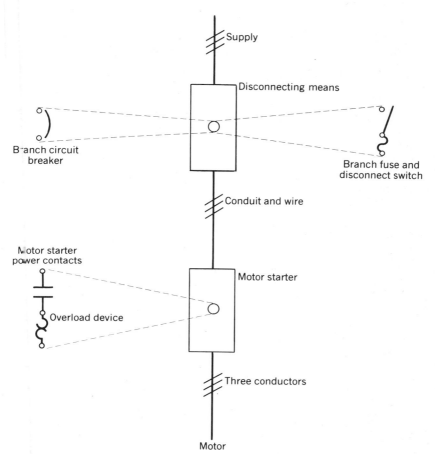

Supply

Disconnecting means

Branch circuit
breaker

Branch fuse and
disconnect switch

Conduit and wire

Motor starter
power contacts

Motor starter

Overload device

Three conductors

Motor

FIGURE 8-22

Single-Line Drawing of the Power Circuit for a Single Motor Installation Started by the Primary Resistance method.

2.   Overload protection.

3.   Conductor sizes.

Overcurrent Protection

The branch-circuit overcurrent device is to be rated at not more than the percentage of the FLA rating of the motor shown in Table 8-1.

Check this statement with the prepared table in the electrical code book having jurisdiction in your particular area. The

TABLE 8-1

Canadian Electrical Code, Part 1

(TABLE 29)

(See Rules 28-200, 28-204, 28-208, and 28-210*)

RATING OR SETTING OF OVERCURRENT DEVICES FOR THE PROTECTION OF MOTOR BRANCH CIRCUITS

Type of Motor	Percent of Full-Load Current		
	Maximum Fuse Rating		Maximum Setting Time-Limit Type Circuit Breaker
	Time-Delay** "D" Fuses	Non-Time Delay	
Alternating Current			
Single-Phase, all types	175	300	250
Squirrel-Cage and Synchronous: Full-Voltage, Resistor and Reactor Starting	175	300	250
Auto-Transformer Starting:			
Not more than 30 A	175	250	200
More than 30 A	175	200	200
Wound Rotor	150	150	150
Direct Current	150	150	150

* (Except as permitted in Table 26 where 15-A overcurrent protection for motor branch-circuit conductors exceeds the values specified in the above table.)

** *Time-delay "D" fuses are those referred to in Rule 14-200.[a]*

NOTES: (1) *The ratings of fuses for the protection of motor branch circuits as given in Table 26 are based on fuse ratings appearing in the table above, which also specifies the maximum settings of circuit breakers for the protection of motor branch circuits.*

(2) *Synchronous motors of the low-torque low-speed type (usually 450 rpm, or lower) such as are used to drive reciprocating compressors, pumps, etc., and which start up unloaded, do not require a fuse rating or circuit-breaker setting in excess of 200% of full-load current.*

(3) *For the use of instantaneous trip (magnetic only) circuit interrupters in motor branch circuits see Rule 28-210.[a]*

[a] The (Table 29) and the Rule numbers mentioned refer to the Canadian Electrical Code. Similar comments will be listed in the National Electrical Code. With permission of Canadian Standards Association, this table is reproduced from CSA Standard C22.1-1982, Canadian Electrical Code, Part 1 (14th edition), which is copyrighted by CSA and copies may be purchased from CSA, 178 Rexdale Blvd., Rexdale, Ontario M9W 1R3.

prepared table will indicate the nearest size fuse or circuit breaker available.

Overload Protection

When selecting the correct overload device, obtain the full-load current from the motor nameplate. Apply the FLA to the overload chart, which is located on the cover or door of the motor starter, and select the proper overload device by catalogue number. Keep in mind the service factor of the motor. If the service factor (SF) is 1.0, the maximum value of the overload device (in amperes) will be 115% of the FLA. With a service factor of 1.15, the maximum value of the overload device will be 125% of the FLA.

Conductor Sizes

Refer to Figure 8-1. It should be noted that with a primary resistance installation, only three power circuit conductors are required.

Refer to the Figure 8-22. The minimum allowable ampacity of the power circuit conductors will be 125% of the FLA of the motor. If for example, the nameplate FLA indicated 100 A, the minimum allowable ampacity of the conductors would be 100 A × 125% = 125 A. (For standardization, 90° C copper conductors will be used in all examples.) The electrical code will require a minimum of a conductor no. 1.

REVIEW EXERCISES

8-1 What is meant by primary resistance starting?

8-2 What part of a squirrel-cage motor is the primary circuit?

8-3 Is this method of starting reduced voltage, or reduced current starting?

8-4 How are the overload devices selected?

8-5 Explain two-stage primary resistance starting.

8-6 Explain four-stage primary resistance starting.

8-7 Is primary resistance starting open or closed transition?

8-8 How is the ampacity of the power circuit conductors determined?

8-9 How is closed-circuit transition achieved by this method of starting?

8-10 Discuss the operation of the compression disc resistor.

8-11 How may sets of resistors are used to provide three-stage, primary resistance starting?

8-12 Describe the operation of a pneumatic time-delay relay.

8-13 What is the function of an overload device?

8-14 What is the purpose of installing overcurrent devices in a power circuit?

8-15 Is it permissible to shunt the overload device out of the circuit during the starting period?

PROBLEMS

The problems are divided into the following three categories:

Part A The application of the electrical code.
Part B Troubleshooting of motor control circuits.
Part C Multiple-choice questions on the combined topics.

Answers and solutions for the odd-numbered problems will be located at the back of this text. Answers and solutions for the even-numbered problems will be contained in the Instructor's Manual.

PART A
ELECTRICAL CODE

Instruction: When answering the following problems, keep in mind that

1. The starting method will be primary resistance.
2. The overload device will be the maximum value in amperes.
3. The conductors will be R90 X-Link CU.

Refer to Figure 8-22.

1. Calculate:
   A. The minimum conductor ampacity for the power circuit and

   _____

   _____

   B. Select the minimum size conductor for the power circuit for an installation when the FLA of the motor is 60 A.

   _____

   _____

   _____

2. The FLA of a squirrel-cage motor is 80 A. Calculate the maximum value of the overload device when the service factor of the motor is 1.0.

   _____

3. The FLA of a squirrel-cage motor is 41 A. Calculate the maximum value of the overload device when the service factor of the motor is 1.15.

   _____

4. Select the maximum allowable rating for a time-delay fuse to be installed as the branch-circuit overcurrent device when the FLA of the squirrel-cage motor is 34 A.

   _____

   _____

PART B
TROUBLESHOOTING

Instructions: For this exercise, a fault will be stated. Study the designated circuit to determine how the circuit will operate.

1. See Figure 8-2. The N/O S contact will not close. Press the start button. Discuss the results.

   _____

   _____

   _____

   _____

2. See Figure 8-2. The TR–TC contact is defective and will not close. Press the start button. Discuss the results.

   _____

   _____

   _____

   _____

   _____

3. See Figure 8-3. The N/C R contact will not open when the R coil is energized. Discuss the results.

   _____

   _____

   _____

   _____

4. See Figure 8-3. The N/O R contact will not close when the R coil is energized. Discuss the results.

   _____

   _____

   _____

5. See Figure 8-4. By accident, the three R power contacts closed before the three S power contacts. Discuss the results.

   _____

   _____

   _____

6. See Figure 8-5. The N/C R contact in series with the S coil is broken and remains open. Press the start button.

Discuss the results.

_____

_____

_____

_____

7. See Figure 8-5. The TR–TC contact is damaged and will not close. Press the start button. Discuss the results.

_____

_____

_____

8. See Figure 8-6. The TSS–TC contact is damaged and will not close. Press the start button. Discuss the results.

_____

_____

_____

_____

_____

_____

PART C
MULTIPLE-CHOICE QUESTIONS

Instructions:

A. The electrical code may be used to assist you in solving the following problems.

B. Select an answer from a, b, c, or d.

C. Answer all questions, based on the drawing in Figure 8-23.

1. The method of starting is:

   A. Secondary resistance.

   B. Primary resistance.

    C. Autotransformer.

    D. Part-winding.

2. The circuit provides:

    A. Three-stage starting.

    B. Two-step starting.

    C. Single stage starting.

    D. Full voltage starting.

3. The coil sequence is:

A.	**B.**
P–IN	$\left.\begin{array}{l}\text{P}\\\text{TR1}\end{array}\right\}$–IN
TR1–IN	
R1–IN	R1–IN
TR2–IN	TR2–IN
R2–IN	R2–IN
TR1–OUT	TR1–OUT

C.	**D.**
$\left.\begin{array}{l}\text{P}\\\text{TR1}\end{array}\right\}$–IN	$\left.\begin{array}{l}\text{P}\\\text{TR1}\end{array}\right\}$–IN
R1–IN	R1–IN
TR2–IN	TR2–IN
R2–IN	R2–IN
	TR2–OUT

4. If the overload contact should open, the motor will stop. Press the start button without resetting the overload device. What will happen?

    A. The P coil will become energized.

    B. The TR1 coil will become energized.

    C. The P and TR1 coils will become energized.

    D. Nothing.

5. If the overcurrent device for this starting method is a time-limit circuit breaker, the maximum rating shall be:

    A. 125% of FLA of the motor.

    B. 175% of FLA of the motor.

    C. 250% of FLA of the motor.

    D. 300% of FLA of the motor.

6. The following method is used to select the overload device:

    A. FLA of the motor divided by 1.73, applied to the electrical code book.

    B. FLA × 250%.

    C.   FLA × 1.73, applied to the chart in the motor starter.

    D.   FLA of the motor, applied to the chart in the motor starter, keeping in mind the service factor.

7.   When the N/C R2 contact opens, what is the next sequence of events?

    A.   The R1 N/O contact closes.

    B.   The R1 coil seals in.

    C.   The R2 coil seals in.

    D.   The TR1 coil deenergizes.

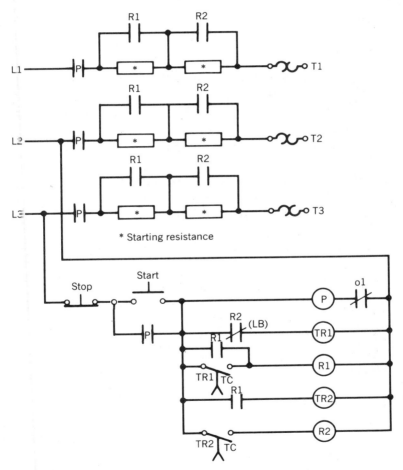

FIGURE 8-23

8. When the start-up sequence has been completed and the motor is running on line voltage, how many coils are energized?
    A. 5.
    B. 3.
    C. 4.
    D. 2.

9. Applying the same data as in no. 8, how many coils are deenergized?
    A. 3.
    B. 2.
    C. 4.
    D. 1.

10. With the overload contact open, press the start button and hold beyond the time-delay settings for both timers. What will happen?
    A. The P coil energizes.
    B. The coil sequence will be TR1–IN, R1–IN, TR2–IN, R2–IN.
    C. The coil sequence will be TR1–IN, R1–IN, TR2–IN, R2–IN, TR1–OUT.
    D. The motor will start.

SECTION 9

WYE-DELTA STARTING

INTRODUCTION TO WYE-DELTA STARTING

The preceding sections have dealt with full voltage, reduced voltage, and reduced current starting methods. This section will study another reduced current starting method. The wye-delta method of starting has proven to be an excellent method for restricting starting current, having been used in Europe for many years prior to its acceptance in North America.

This text does not intend to present the pros and cons for any particular starting method, but rather to teach how to recognize, understand, install, and troubleshoot the various motor starting methods.

Wye-delta motor starting is very often referred to as star-delta starting. The term wye or star is in reference to the starting configuration of the stator winding of a three-phase squirrel-cage motor. The term delta is in reference to the running configuration of the stator winding of the motor. The prerequisite for wye-delta motor starting is a three-phase delta connected motor with all six leads extended to the motor connection terminal box.

Figure 9-1 shows the stator windings and all lead identification for a wye-delta motor. The motor is designed to start with

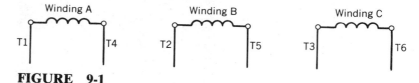

FIGURE 9-1

Lead Identification for a Wye-Delta Motor.

the three stator windings connected in the star configuration (see Figure 9-2). It is also designed to run with the stator windings connected in the delta configuration (see Figure 9-3).

To reverse the direction of rotation of a motor used in the wye-delta starting method, interchange any two-phase conductors on the line side of the motor starter.

In Section 6, we studied autotransformer starting, using autotransformers to supply a lower starting voltage to the stator windings, which reduced the starting current and, in turn, reduced the starting torque. With the wye-delta starting

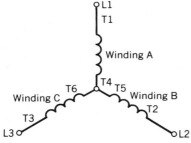

FIGURE 9-2

Starting Configuration for a Wye-Delta Started Motor.

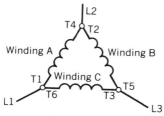

FIGURE 9-3

Running Configuration for a Wye-Delta Started Motor.

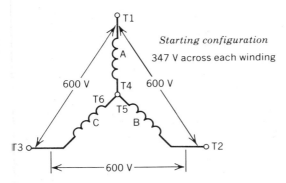

FIGURE 9-4

Voltage Relationship for the Wye Configuration.

method, neither autotransformers nor resistors are used to reduce the starting voltage to the stator windings.

The stator windings are designed to run in the delta configuration on line voltage. When the stator windings are connected in the wye configuration and line voltage is applied to the T1, T2, and T3 leads, the windings that are designed for line voltage receive approximately 58% of line voltage. This results in a lower starting current. Refer to Figures 9-4 and 9-5. Assume the supply voltage to be three-phase, 600 V.

OPERATION OF WYE-DELTA MOTOR STARTING

Figure 9-4 shows the starting position with approximately 58% of line voltage applied to the stator windings. The starting

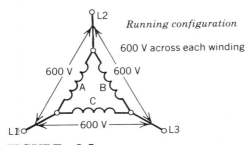

FIGURE 9-5

Voltage Relationship for the Delta Configuration.

current is considerably lower than it would be if line voltage were applied to the delta configuration at start. Following a preset time delay, the star point is opened, and the motor is reconnected manually or automatically to the delta configuration (Figure 9-5). The stator winding is now receiving line voltage (600 V), increasing the torque, and the motor is able to run as a three-phase, delta-connected squirrel-cage motor.

The motor lead identifications of T1, T2, T3, T4, T5, and T6 must be connected to the matching terminations in the motor starter. This is of the utmost importance.

TRANSITION

Transition, as described in Section 5, may be either open or closed. Wye-delta is primarily open transition, but with the

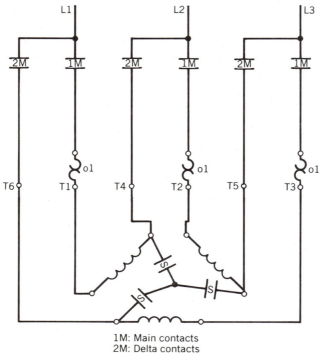

1M: Main contacts
2M: Delta contacts
S: Star contacts

FIGURE   9-6

Power Circuit Diagram of a Wye-Delta, Open Transition Motor Starter with Motor Stopped.

addition of factory-installed resistors, closed transition may be achieved. Figure 5-9 shows how a spike may occur when a motor is started open transition.

The power circuits in Figures 9-6 through 9-10 explain the circuit operation for open transition wye-delta. Figure 9-6 illustrates the power circuit as it would appear when the motor is not operating, and all power contacts are in the open position. The power circuit as depicted in Figure 9-7 shows the 1M and S contacts closed, with the 2M contacts in the open position.

Following the power circuit from L1, L2, and L3, we will notice that the stator windings are connected in a wye configuration similar to that in Figure 9-4. If we assume that the supply voltage is 600 V, three-phase, the voltage across each winding

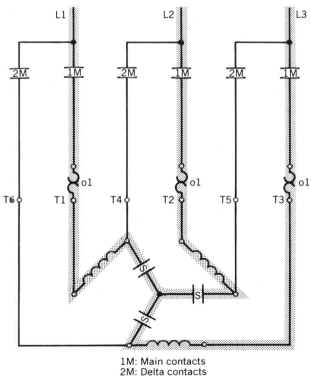

1M: Main contacts
2M: Delta contacts
S: Star contacts

FIGURE 9-7

The Shaded Portion of the Power Circuit Shows the Beginning of the Starting Sequence.

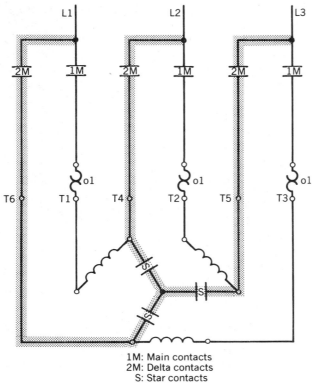

1M: Main contacts
2M: Delta contacts
S: Star contacts

FIGURE 9-8

The Shaded Portion Shows the Possible Short Circuit, as Mentioned in the Text.

will be 347 V. The lower starting voltage across each winding produces a lower starting current.

The current in Figure 9-8 is the short-circuit that would occur if the 2M and S contacts were to close at the same time. Safety contacts and interlocking devices are provided that will prevent this from happening. The shaded portion of the power circuit in Figure 9-9 shows the incompleted circuit following the opening of the S contacts and prior to the closing of the 2M contacts. The motor is no longer connected to the line. Following the closing of the 2M power contacts, the motor is delta-connected and running with line voltage across each winding. Figure 9-10 illustrates the completed circuit.

The control circuit in Figure 9-11 will cause the power circuit to operate as described in Figures 9-6 through 9-10.

When we press the start button in Figure 9-11, a circuit is completed for the 1M, TR, and S coils, which closes the 1M and S power contacts. At the same instant, the N/O 1M contact closes maintaining the circuit, and the N/C S electrical interlock opens.

After a preset time delay, the TR–TO contact opens, de-energizing the S coil, and the TR–TC contact closes. When the N/C S electrical interlock recloses, the 2M coil becomes energized, closing the 2M power contacts connecting the motor windings in the delta configuration.

Compare the power and control circuits shown in Figure 9-12 with the combined circuits in Figures 9-6 and 9-11. It will be noted that the timed contacts in Figure 9-11 are released by energizing a TR coil, while in Figure 9-12, the timed contacts are released by energizing the 1M coil. The two circuits are

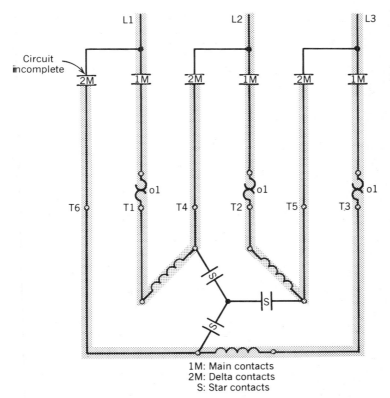

1M: Main contacts
2M: Delta contacts
S: Star contacts

FIGURE 9-9

The Shaded Portion of the Power Circuit Illustrates the Open Transition Part of the Starting Sequence.

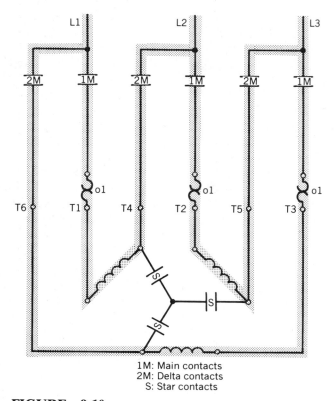

1M: Main contacts
2M: Delta contacts
S: Star contacts

FIGURE   9-10

The Shaded Portion of the Power Circuit Shows the Motor in the Running Operation.

otherwise the same. Power contact identification may differ, but the power circuit operation remains the same. The N/C 2M contact in the control circuit of Figure 9-12 is an extra safeguard, as the S and 2M coils must not be energized at the same time.

The power circuits in Figures 9-12 and 9-13 use the same identification: 1M, 2M, and S. Allen-Bradley uses 3–S contacts in the power circuit, while Square D uses 2–S contacts to perform the same function. Both power circuits will operate in the same manner. Allen-Bradley has chosen to trigger the time-delay contacts, while Square D employs a timer coil (TR) to release the time-delay contact.

Each motor control manufacturer has the right to select the method best-suited to their equipment, and we as tradespeo-

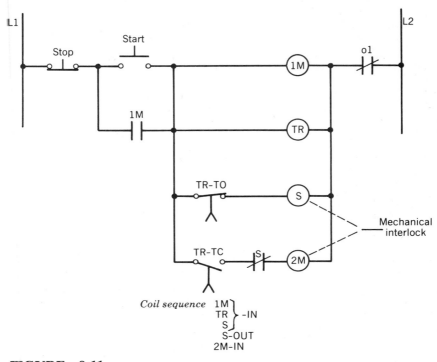

FIGURE   9-11

Open Transition Control Circuit for Wye-Delta Starting.

ple should not be overly concerned with the question of which
is better. Our responsibility is to install and service the equipment for which we are responsible.

Both circuits will perform the operation shown in Figures
9-7 through 9-10.

Figures 9-6 to 9-13 dealt with open transition. Wye-delta
motor starters may, at the manufacturing level, be equipped
with an additional three-pole contactor and a set of resistors.
The resistors most often are variable; this allows the changeover spike to be controlled to the lowest level possible.

Closed transition is accomplished by connecting resistors
across the delta power contacts for a split second at the
changeover period. The power circuits in Figures 9-14
through 9-19 will serve to explain the circuit operation for
closed transition wye-delta.

Figure 9-14 shows the power circuit as it would appear
when the motor is not operating and all power contacts are in

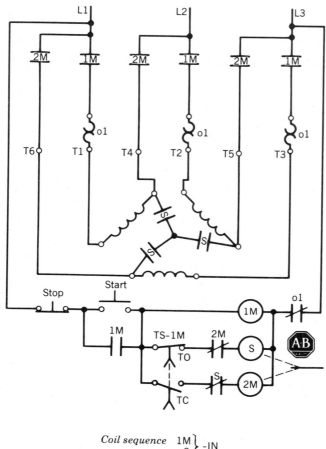

FIGURE 9-12

Wye-Delta Motor Starting. Power and Control Circuits. Open Transition (example one).

the open position. The power circuit in Figure 9-15 depicts the 1M and 1S contacts closed, with the 2M and 2S contacts in the open position.

Following the power circuit from L1, L2, and L3, we will notice the stator windings are connected in a wye configuration similar to that in Figure 9-4. If we assume that the supply voltage is 600 V, three-phase, the voltage across each winding will be 347 V. The lower starting voltage across each winding will produce a lower starting current.

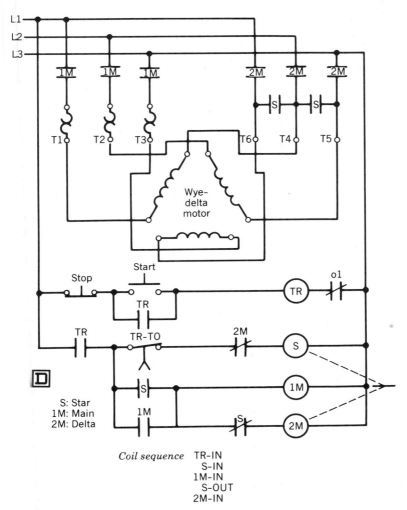

FIGURE 9-13

Wye-Delta Motor Starting. Power and Control Circuits. Open Transition (example two).

Following a preset time delay, the 2S contacts will close, connecting the resistors across the 2M contacts as illustrated in Figure 9-16, prior to opening of the S contacts. The next step is to open the S contacts. For an instant, the motor is connected in the delta configuration with resistors in series with each winding. Figure 9-17 shows the circuit at this point.

Following the closing of the 2M contacts, the resistors are

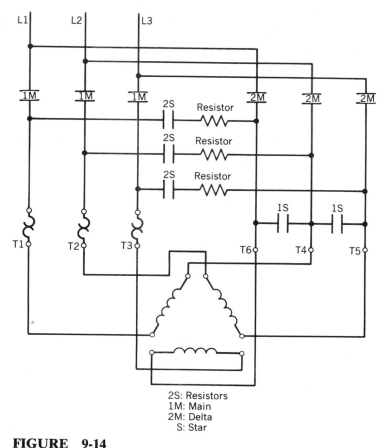

FIGURE 9-14

Power Circuit Diagram for a Wye-Delta, Closed Transition Motor Starter with Motor Stopped.

shorted out of the circuit, and each motor winding is receiving line voltage. Figure 9-18 represents the circuit at this stage of the operation. The final step in the starting operation is shown in Figure 9-19. The R contacts have opened, removing the resistors from the circuit.

The control circuits in Figures 9-20 and 9-21 could be used to operate the power circuit shown in Figure 9-14. Compare the two. The circuits differ in layout and design, but both will perform the required operation.

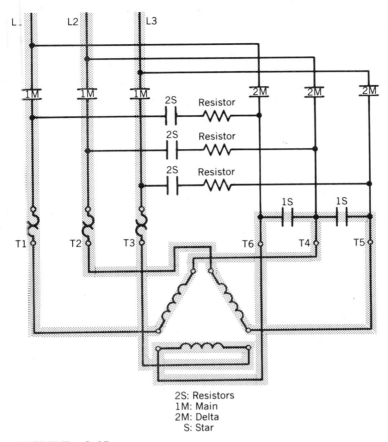

2S: Resistors
1M: Main
2M: Delta
S: Star

FIGURE 9-15

The Shaded Portion of the Power Circuit Shows the Beginning of the Starting Sequence.

POWER CIRCUIT CALCULATION

Single Motor Installation

To design the power circuit for a single motor installation, the following calculations must be understood:

1. Overcurrent protection.
2. Overload protection.

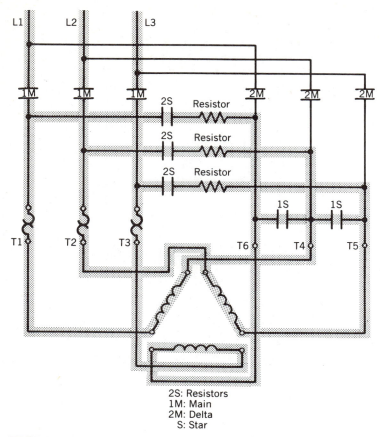

2S: Resistors
1M: Main
2M: Delta
S: Star

FIGURE 9-16

The Shaded Portion of the Power Circuit Shows the First Part of the Transition.

3. Conductor sizes
   A. To the supply side of the starter.
   B. On the load side of the starter.

Overcurrent Protection

The branch-circuit overcurrent device is to be rated not more than the percentage shown in Table 9-1 of the FLA rating of the motor. Check this statement with the prepared table in the electrical code book having jurisdiction in your particular area. The prepared table will indicate the nearest size fuse or circuit breaker available.

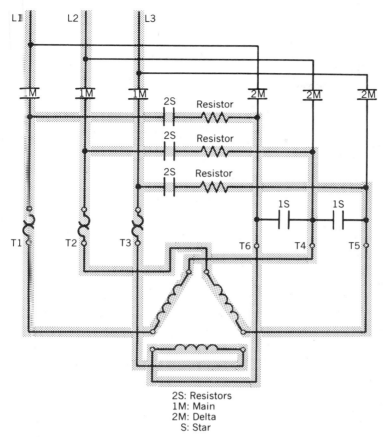

2S: Resistors
1M: Main
2M: Delta
S: Star

FIGURE   9-17

The Shaded Portion of the Power Circuit Shows the Next Part of the Transition.

Overload Protection

When selecting the correct overload device, it must be determined what part of the motor nameplate current actually passes through the overload heater during the running period. The motor nameplate current is line current, not phase current. When a motor starts on line voltage and runs on line voltage, the overload heater is selected according to line current.

Refer to Figure 9-6. It will be noticed that the overload heater (relay) is connected to monitor phase current, and not line current.

$$\frac{Line\ current}{\sqrt{3}} = Phase\ current$$

The stator winding is connected in the delta configuration during the running period. Study Figure 9-22.

How to Select the Overload Device

When selecting the overload device, obtain the full-load current from the motor nameplate. The current for each winding will be

Full load current divided by $\sqrt{3}$

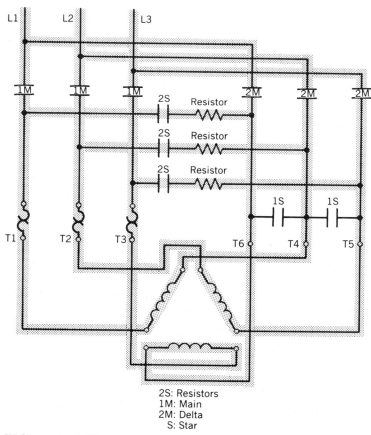

2S: Resistors
1M: Main
2M: Delta
S: Star

FIGURE 9-18

The Shaded Portion of the Power Circuit Shows the Third Part of the Transition.

If, for example, the FLA is 50 A, each winding will be 50 A divided by $\sqrt{3}$ = 28.87 A. Apply this figure (28.87 A) to the overload chart located on the cover or door of the motor starter and select the proper overload device by catalogue number. Keep in mind the service factor of the motor.

Conductor Sizes

To select the conductors to the supply side of the motor starter, the minimum ampacity of the conductors must be 125% of the motor nameplate current. To select the conductors from the motor starter to the motor terminals, the mini-

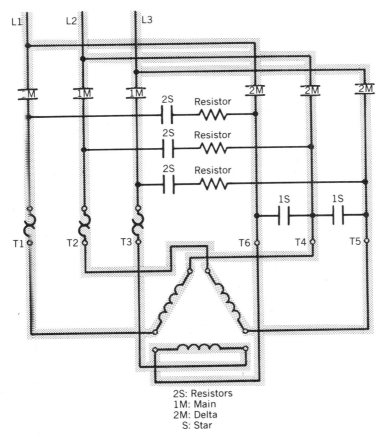

2S: Resistors
1M: Main
2M: Delta
S: Star

FIGURE 9-19

The Shaded Portion of the Power Circuit Shows the Motor in Running Operation.

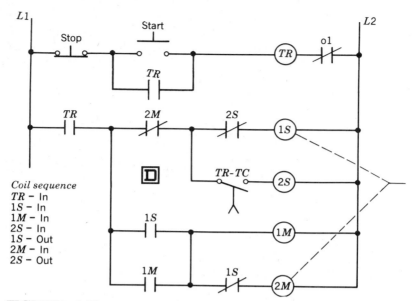

FIGURE 9-20

Closed Transition Control Circuit for Wye-Delta Starting.

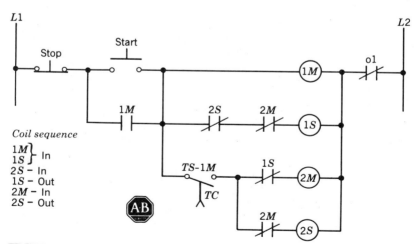

FIGURE 9-21

Closed Transition Control Circuit for Wye-Delta Starting.

TABLE 9-1

Canadian Electrical Code, Part 1

(TABLE 29)

(See Rules 28-200, 28-204, 28-208, and 28-210*)

RATING OR SETTING OF OVERCURRENT DEVICES FOR THE PROTECTION OF MOTOR BRANCH CIRCUITS

Type of Motor	Percent of Full-Load Current		
	Maximum Fuse Rating		**Maximum Setting Time-Limit Type Circuit Breaker**
	Time-Delay** "D" Fuses	**Non-Time Delay**	
Alternating Current			
Single-Phase, all types	175	300	250
Squirrel-Cage and Synchronous: *Full-Voltage,* Resistor and Reactor Starting	175	300	250
Autotransformer Starting[b]:			
Not more than 30 A	<u>175</u>	<u>250</u>	<u>200</u>
More than 30 A	<u>175</u>	<u>200</u>	<u>200</u>
Wound Rotor	150	150	150
Direct Current	150	150	150

* (Except as permitted in Table 26 where 15-A overcurrent protection for motor branch-circuit conductors exceeds the values specified in the above Table.)

** *Time delay "D" fuses are those referred to in Rule 14-200.*[a]

NOTES: (1) *The ratings of fuses for the protection of motor branch circuits as given in Table 26 are based on fuse ratings appearing in the table above, which also specifies the maximum settings of circuit breakers for the protection of motor branch circuits.*

(2) *Synchronous motors of the low-torque low-speed type (usually 450 rpm, or lower) such as are used to drive reciprocating compressors, pumps, and which start up unloaded, do not require a fuse rating or circuit-breaker setting in excess of 200% of full-load current.*

(3) *For the use of instantaneous trip (magnetic only) circuit interrupters in motor branch circuits see Rule 28-210.*[a]

[a] Table 29 and the Rule numbers mentioned are contained in the Canadian Electrical Code. Similar comments will be listed in the National Electrical Code. With permission of Canadian Standards Association, this table is reproduced from CSA Standard C22. -1982, Canadian Electrical Code, Part 1 (14th edition), which is copyrighted by CSA and copies may be purchased from CSA, 178 Rexdale Blvd., Rexdale, Ontario M9W 1R3.

[b] Same percentages are used for wye-delta as an autotransformer.

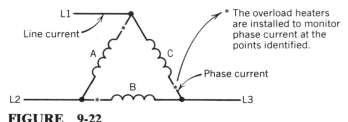

FIGURE 9-22

Line Current vs. Phase Current.

mum ampacity of the conductors must not be less than 125% of the current rating of the winding configuration they serve.

Refer to Figure 9-6. It should be noted that three conductors are required to the supply side of the motor starter, and six conductors from the motor starter to the motor.

Various wiring methods may be used to complete the power circuit installation. The following lessons will provide the student with instruction for three different wiring installations.

POWER CIRCUIT CONDUCTOR CALCULATION

Single Motor Installation

Figure 9-23 depicts an installation using single conduit runs and 90°C copper conductors. To calculate the minimum allowable ampacity of the conductors to the supply side of the motor starter, the nameplate current rating of the motor must be used.

Example: If the nameplate indicates 120 A, the minimum allowable ampacity of the conductors must be 120 A × 125% = 150 A. The electrical code will require a minimum of 1/0.

To calculate the minimum allowable ampacity of the conductors from the motor starter to the motor, the conductors must have a minimum ampacity of 125% of the winding they serve. The conductors from the motor starter to the motor must be sized for phase current plus 25%. Therefore, if the FLA of the motor is 120 A, the conductors must be rated for 120 A divided by $\sqrt{3}$ × 125% = 86.6 A.

It must be remembered that when six conductors are installed in a single raceway, the conductor ampacity must be reduced to 80% of the ampacity shown in the appropriate

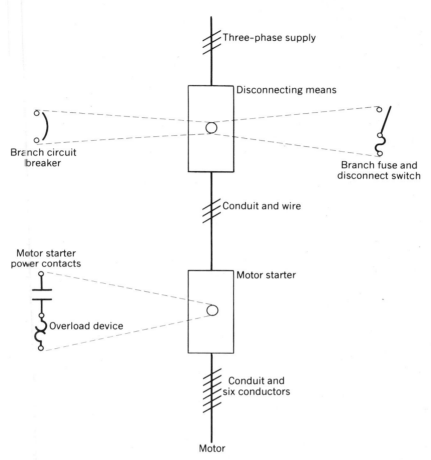

FIGURE 9-23

Single-Line Drawing of the Power Circuit for a Single Motor Installation, Started by the Wye-Delta Method.

table in the electrical code book. What size conductors would be required?

No. 4 90C copper conductors are rated at 90 A. The conductor ampacity with six conductors installed in the same raceway would be 80% of 90 A, or 72 A. The required ampacity is 86.6 A. No. 2 90C copper conductors have a rating of 120 A. Derated to 80% would mean 96 A. Therefore, six no. 2 90C copper conductors would be required from the motor starter to the motor if a single conduit is installed as shown in Figure 9-23.

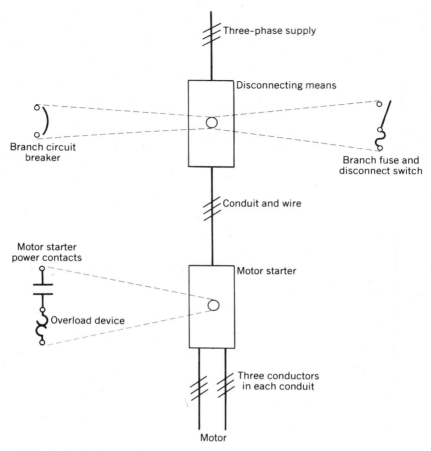

Three-phase supply

Disconnecting means

Branch circuit breaker

Branch fuse and disconnect switch

Conduit and wire

Motor starter power contacts

Motor starter

Overload device

Three conductors in each conduit

Motor

FIGURE   9-24

Single-Line Drawing of the Power Circuit for a Single Motor Installation, Started by the Wye-Delta Method.

Figure 9-24 shows an installation using two conduit runs from the motor starter to the motor. One conduit will contain the T1, T2, and T3 conductors, and the second conduit will contain the T4, T5, and T6 conductors.

Using the data from the previous installation that required conductors rated for at least 86.60 A, we shall determine the conductor size to be no. 4 (90° copper). The conductor ampacity need not be derated as there are only three conductors in each raceway.

Figure 9-25 shows an installation using single conductor

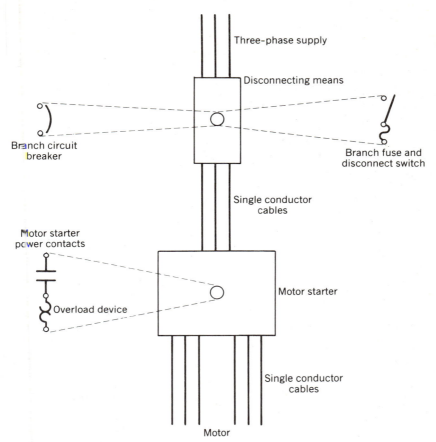

FIGURE 9-25

Single-Line Drawing of the Power Circuit for a Single Motor Installation, Started by the Wye-Delta Method.

cables 90C − CU. To calculate the minimum ampacity of the conductors to the supply side of the motor starter, the nameplate current rating of the motor must be used.

Example: If the FLA were 200 A, the minimum ampacity of the conductors would be 200 A × 125% = 250 A. Therefore, 2/0 conductors would be required.

The minimum ampacity of the conductors from the motor starter to the motor is determined from phase current times 125%. Therefore, no. 3 conductors would be required.

$$(200 \text{ A divided by } \sqrt{3}) \times 125\% = 144.34 \text{ A.}$$

REVIEW EXERCISES

9-1   Explain the wye configuration as related to a three-phase motor.

9-2   Explain the delta configuration as related to a three-phase motor.

9-3   Is the wye-delta method of starting reduced current, or reduced voltage starting?

9-4   How are the overload devices selected?

9-5   Explain open transition.

9-6   Explain closed transition.

9-7   How is closed transition accomplished with this method of starting?

9-8   Explain line current.

9-9   Explain phase current.

9-10   Explain line voltage.

9-11   Explain phase voltage.

9-12   Explain the function of the mechanical interlock in a wye-delta motor starter.

9-13   Do the overload heaters monitor line current or phase current?

9-14   How many power circuit conductors are required from the motor starter to the motor for this method of starting?

9-15   How is the minimum ampacity of the power circuit conductors from the motor starter to the motor determined?

PROBLEMS

The problems are divided into the following three categories:

Part A   The application of the electrical code.
Part B   Troubleshooting of motor control circuits.
Part C   Multiple-choice questions on the combined topics.

Answers and solutions for the odd-numbered problems will be located at the back of this text. Answers and solutions for the even-numbered problems will be contained in the Instructor's Manual.

PART A
ELECTRICAL CODE

Instructions: When answering the following problems, keep in mind:

1. The starting method will be wye-delta.
2. The overload device will be the maximum value in amperes.
3. The conductors will be R90 X-Link CU.
4. The motor FLA will be 68 A.

Refer to Figure 9-24.

1. Calculate:
   A. The minimum conductor ampacity for the power circuit and

   _____

   _____

   B. Select the minimum size conductor for the power circuit from the motor starter to the motor.

   _____

2. Calculate the maximum value of the overload device if the motor service factor is 1.0.

   _____

   _____

3. Select the maximum allowable rating for a time-delay fuse to be installed as the branch-circuit overcurrent device.

   _____

4. Select the minimum size conductor for the power circuit to the line side of the motor starter.

   _____

   _____

PART B
TROUBLESHOOTING

Instructions: For this exercise, a fault will be stated. Study the designated circuit to determine how the circuit will operate.

1. See Figure 9-11. The conductor on the left side of the TR coil has dropped off the TR coil terminal. Press the start button. What will happen?

    _____

    _____

2. See Figure 9-11. The S N/C contact is broken and will not close. Press the start button. What will happen?

    _____

    _____

    _____

    _____

3. See Figure 9-11. The TR–TO contact does not open following the preset time delay. Press the start button. What will happen?

    _____

    _____

    _____

4. See Figure 9-12 (control circuit). The N/C M contact is welded shut and will not open. Press the start button. What will happen?

    _____

    _____

5. See Figure 9-13 (control circuit). The TR N/O contact below the stop button is broken and will not close. Press the start button. What will happen?

    _____

    _____

    _____

6. See Figure 9-20. The 2M N/C contact is defective and will not open after the 2M coil has been energized. How will the circuit operate?

_____

_____

_____

_____

PART C
MULTIPLE-CHOICE QUESTIONS

Instructions:

A. The electrical code book may be used to assist you in solving the following problems.

B. Select answers from a, b, c, or d.

C. Answer all questions, based on the drawings in Figures 9-26 and 9-27.

1. The starting method is:
   A. Wye-delta.
   B. Secondary resistance.
   C. Primary resistance.
   D. Full voltage.

2. The type of transition is:
   A. Overcurrent.
   B. Closed.
   C. Open.
   D. Overload.

3. The control circuit provides:
   A. No voltage release.
   B. No field release.
   C. No voltage protection.
   D. Undervoltage release.

4. Press the start button. Before the TR–TC contact closes, the following power circuit contacts are closed.

    A. 1S and 1M.

    B. 1S and 2S.

    C. 1M and 2M.

    D. Only 2M.

5. When the motor is running on line voltage, the following power circuit contacts are open:

    A. 1S and 1M.

    B. 1M and 2M.

    C. 1S and 2S.

    D. 2S and 1M.

6. When the motor is running on line voltage, the following power circuit contacts are closed:

    A. 1S and 2S.

    B. 1S and 1M.

    C. 1M and 2M.

    D. 2S and 1M.

7. In the starting position, the motor stator windings are connected:

    A. Open delta.

    B. Star (wye).

    C. Delta.

    D. Open.

8. In the running position, the motor windings are connected:

    A. Open delta.

    B. Star (wye).

    C. Delta.

    D. Open.

9. The overload heaters are selected by which method?

    A. FLA divided by $\sqrt{3}$, applied to the chart in the motor starter.

    B. FLA × 200%.

    C. FLA divided by 2, applied to the electrical code book.

    D. FLA × 300%.

10. Select the correct coil sequence.

A. TR–IN
1S–IN
1M–IN
2S–IN
1S–OUT
2M–IN
2S–OUT

B. TR
1S $\Big\}$–IN
1M–IN
2S–IN
1S–OUT
2M–IN
2S–OUT

C. 1S–IN
TR–IN
1M–IN
2S–OUT
1S–OUT
2M–IN

D. TR–IN
1S–IN
1M
2S $\Big\}$–IN
1S–OUT
2M
2S $\Big\}$–OUT

11. If the N/C 2M contact in the control circuit did not open as it should, the 2S coil would remain energized. What effect would this have on the operation of the motor?

A. Trip the overload current.

B. Decrease the torque of the motor.

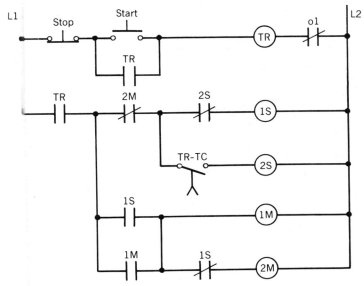

FIGURE 9-26

    C. None.

    D. Increases coil current.

12. A malfunction in the motor starter during the start-up period has allowed the 2M and the 1S power contacts to be closed at the same time. What would happen?

    A. It would have no effect on the circuit.

    B. It would heat up the resistors.

    C. The overload heater would trip.

    D. It would cause a short circuit.

13. The maximum setting of the overcurrent protection (time-limit circuit breaker) installed to protect the power circuit conductors would be:

    A. 300% of FLA.

    B. 200% of FLA.

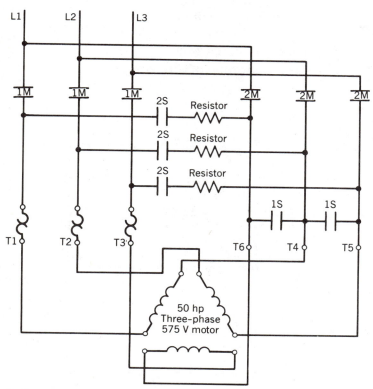

FIGURE 9-27

    C. FLA divided by 2.

    D. FLA × 150%.

14. When the TR–TC contact closes, the next sequence in the starting operation will be:

    A. 1S coil is deenergized.

    B. 2S coil is energized.

    C. 2M coil is energized.

    D. 2S N/C contact opens.

15. Under an overload condition, an overload contact opens. What is the next sequence in the operation?

    A. TR coil is deenergized.

    B. TR contacts switch positions.

    C. 1S coil is deenergized.

    D. Nothing noticeable happens.

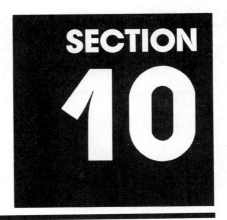

SECTION 10

SECONDARY RESISTANCE STARTING

INTRODUCTION TO SECONDARY RESISTANCE STARTING

The wound rotor induction motor is the least understood motor of the polyphase induction motor line. This text is the study of the control for the wound rotor induction motor, and not the motor design itself.

The wound rotor induction motor is used when controlling starting torque and speed is a requirement. Indiscriminate matching of starting resistors and wound rotor induction motors should be discouraged. Table 10-1 is a typical request form to properly advise control engineers and designers as to the starting and running requirements for a particular installation.

When starting a wound rotor induction motor, the stator is supplied with line voltage, while external resistance units are connected in the rotor circuit. This starting method is reduced current starting, and not reduced voltage starting.

PRIMARY CIRCUIT VERSUS SECONDARY CIRCUIT

The primary circuit of a wound rotor induction motor is the stator winding, while the rotor winding is referred to as the

TABLE 10-1

Request for Wound Rotor Induction Motor Data, Allen-Bradley Company

OFFICE_____ MEMO_____ DATE_____

CUSTOMER_____

C.O. #_____F.O. #_____P.W. #_____

H.P._____LINE VOLTS_____PHASE_____HERTZ_____

PRIMARY F.L.C._____ SERVICE FACTOR_____

SECONDARY F.L.C._____SECONDARY VOLTS_____PHASE_____

TYPE OF RESISTOR. CAST GRID_____NON-BREAKABLE_____

NUMBER OF ACCELERATING POINTS OR SPEED REGULATING POINTS

_____

DESCRIBE ALL SPECIAL REQUIREMENTS IN DETAIL._____

_____

_____

_____

FOR STARTING DUTY

RESISTOR CLASS_____ACCELERATING TIME_____

FREQUENCY OF STARTING_____

% OF FULL LOAD TORQUE ON FIRST POINT_____

DESCRIBE LOAD DRIVEN BY MOTOR_____

_____

_____

FOR REGULATING DUTY

TYPE OF LOAD. FAN DUTY_____MACHINE DUTY_____OTHER_____

TABLE   10-1

IF OTHER DESCRIBE OR SUPPLY SPEED-TORQUE CURVE OF LOAD.

_____

_____

_____

_____

IS THE MOTOR FULLY LOADED AT FULL SPEED?

GIVE PERCENT._____%

WHAT SPEED REDUCTION IS REQUIRED?_____%

WHAT PERCENT OF MOTOR FULL LOAD TORQUE DOES THE LOAD

REQUIRE AT THAT SPEED?_____%

secondary circuit; hence, the term "secondary resistance" starting.

WOUND ROTOR INDUCTION MOTOR CIRCUITS

Wound rotor induction motors have star-connected stator and rotor windings. The stator winding is similar to the winding of a squirrel-cage motor. The star-connected winding of the rotor terminates at three slip-rings located at one end of the rotor.

Figure 10-1 shows a basic power circuit used to supply line voltage to the stator windings and external resistors connected to the rotor by means of a set of brushes.

RESISTANCE UNITS

The resistance units illustrated in Figure 10-2 are located in an enclosure separate from the motor and wired with conductors having insulation rated for the temperature encountered from the resistance units.

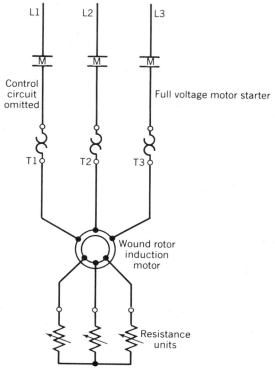

FIGURE 10-1

Basic Wound Rotor Induction Motor Power Circuit.

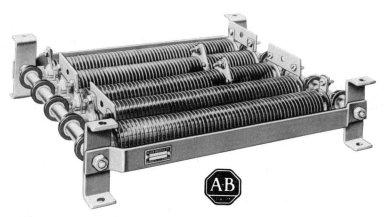

FIGURE 10-2

Bank of Resistors Used in Secondary Resistance Rotor Circuits.

IMPORTANCE OF THE SECONDARY CIRCUIT

It is imperative that when full voltage is applied to the stator at start, the full external resistance be connected to the rotor circuit. Why?

1. The motor will not start if the rotor circuit is open.
2. The starting current will be extremely high if the resistors were bypassed.

The control circuit for the magnetic motor starter in Figure 10-1 is depicted in Figure 10-3.

Circuit Operation, Figure 10-3

When the handle for the controller is at maximum resistance, the resistor interlock is closed. Pressing the start button will complete the control circuit for the M coil. The control circuit contact M bypasses both the resistor interlock and the start button, maintaining the circuit for the M coil.

When the M power contacts close, line voltage is applied to the stator winding, and full resistance is connected to the rotor circuit. Rotating the manual control decreases the resistance, increasing motor speed.

Most secondary resistance installations are semiautomatic or fully automatic, as opposed to manual starting.

Semiautomatic   A semiautomatic system is one in which an operator presses an initiating device, such as a pushbutton, with the remainder of the operation being automatic.

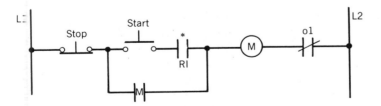

* RI: This contact is referred to as a resistor interlock

FIGURE 10-3

Control Circuit for the Power Circuit Shown in Figure 10-1.

Fully Automatic A fully automatic system is one in which a device starts the sequence of operation without relying on an operator to press a button. An example of such a system would be a two-wire control that operates on a change of temperature or pressure.

Semi- or Fully Automatic Figure 10-1 shows a magnetic motor starter used to supply full voltage to the stator winding. If we added magnetic contactors to the circuit, the installation could be changed to a semiautomatic or a fully automatic starting system.

Multistage or Step Secondary resistance starting, similar to primary resistance starting, may have two, three, four, or more steps or stages.

Secondary resistance starting is closed transition, regardless of the number of stages, or whether the circuit is automatic or manual.

TWO-STAGE STARTING

Circuit Operation, Figure 10-4

Figure 10-4 illustrates a typical two-stage starting circuit for a wound rotor motor. Pressing the start button completes a circuit for the M and TR coils, closing the M (control circuit) contact, which maintains the control circuit. At the same instant, the M power contacts close, connecting the stator winding to line voltage. The TR–TC contact has been released. Following the preset time delay, the TR–TC contact closes, energizing the R coil, closing the R power contacts, bypassing the resistors. The motor continues to run as a squirrel-cage motor. The motor is never off the line during the start-up operation; therefore, the starting method is closed transition.

Figure 10-4 represents a semiautomatic system. Figure 10-5 shows a fully automatic system using the power circuit in Figure 10-4.

The power circuit performs in the same manner as de-

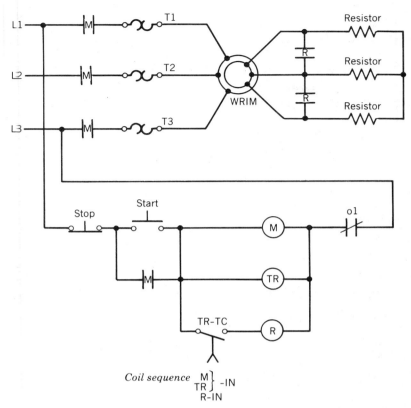

FIGURE 10-4

Two-Stage Secondary Resistance Starting. Power and Control Circuits. NVP Circuit.

scribed for Figure 10-4, even though the control circuit in Figure 10-5 differs. Compare the control circuits illustrated in Figures 10-4 and 10-5. The coil sequence may vary, but the end result is the same. Learn to read, design, and understand various circuits, but do not attempt to memorize any.

Various Two-Stage Control Circuits

Figures 10-6, 10-7, and 10-8 are control circuits designed to prove the point that control circuits may differ, but the operation of the power circuit and the equipment will be the same. Spend time with the previous two-stage control circuits in order to be convinced of this fact. Each uses TR coils to

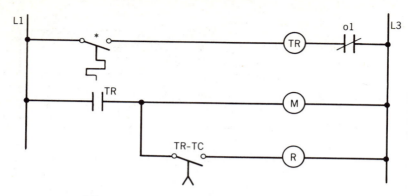

* Temperature control closes on temperature rise.

FIGURE 10-5

Two-Stage Secondary Resistance Starting. Control Circuit. NVR Circuit (example one).

release time-delay contacts. Some motor control companies do not use TR coils in the circuit design at all (refer to Section 5).

Figure 10-9 shows a control circuit employing a timing unit triggered by the M contactor in Figure 10-4.

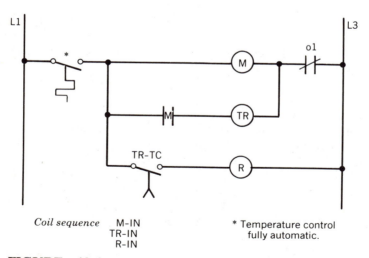

Coil sequence   M–IN
                      TR–IN
                      R–IN

* Temperature control
fully automatic.

FIGURE 10-6

Two-Stage Secondary Resistance Starting. Control Circuit. NVR Circuit (example two).

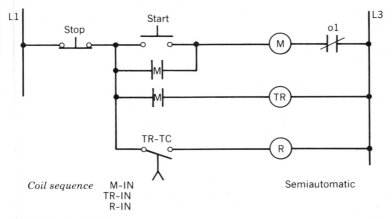

FIGURE 10-7

Two-Stage Secondary Resistance Starting. Control Circuit. NVP Circuit (example one).

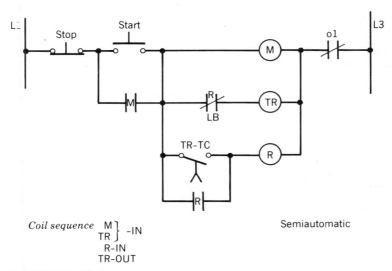

FIGURE 10-8

Two-Stage Secondary Resistance Starting. Control Circuit. NVP Circuit (example two).

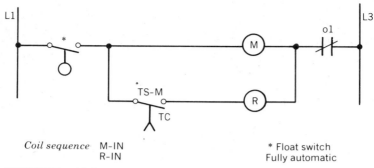

Coil sequence   M-IN
            R-IN

* Float switch
Fully automatic

FIGURE   10-9

Two-Stage Secondary Resistance Starting. Control Circuit. NVR Ci-cuit (using triggered time delay).

THREE-STAGE STARTING

Power Circuit Operation, Figure 10-10

Figure 10-10 is a schematic diagram of the power circuit for a three-stage, nonreversing, secondary resistance starter. The first stage is established when the M contacts close, applying full voltage to the stator winding. Two sets of secondary resistors are connected to the rotor, which restricts the starting current in the supply conductors.

Following a preset time delay, the R1 power contacts close, bypassing one set of resistors, increasing the motor speed. The third and final stage follows the closing of the power contacts R2, which shorts out the second set of resistors, increasing the motor speed to maximum.

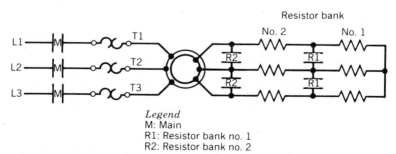

Legend
M: Main
R1: Resistor bank no. 1
R2: Resistor bank no. 2

FIGURE   10-10

Three-Stage Secondary Resistance Starting. Power Circuit.

Various control circuits for three-stage, nonreversing secondary resistance starting are shown in Figures 10-11, 10-12, 10-13, and 10-14. Any one of the control circuits will cause the power circuit to operate as discussed in the power circuit operation for Figure 10-10.

Control Circuit Operation, Figure 10-11

Pressing the start button completes a circuit for the TR1 and M coils, closing the M contact, which maintains the control circuit, and releasing the TR1–TC contact. Following a preset time delay, the TR1–TC contact closes, completing a circuit for the R1 coil, closing the R1 contact, completing a circuit for the TR2 coil. Contact TR2–TC is released, and following a preset time delay, the contact closes, completing a circuit for the R2 coil.

Compare the circuits in Figures 10-10 and 10-11. It will be noted that the circuits are compatible.

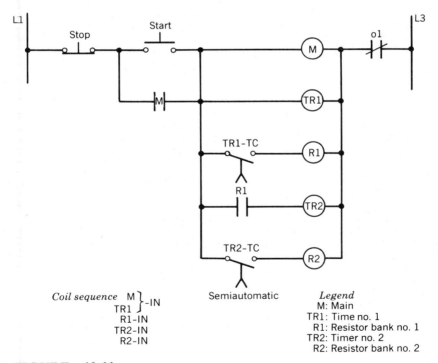

Coil sequence M ⎫
     TR1 ⎬ -IN
     R1-IN
     TR2-IN
     R2-IN

Semiautomatic

Legend
M: Main
TR1: Time no. 1
R1: Resistor bank no. 1
TR2: Timer no. 2
R2: Resistor bank no. 2

FIGURE 10-11

Three-Stage Secondary Resistance Starting, Nonreversing. Control Circuit. NVP Circuit (example one).

Control Circuit Operation, Figure 10-12

When the temperature control closes, a circuit is completed for the M and TR1 coils, releasing the TR1 contact. Following the preset time delay, the TR1–TC contact closes, completing a circuit for the R1 and TR2 coils. TR2–TC contact is released and closes after the preset time delay, completing a circuit for the R2 coil. Contact R1 (N/C) opens, deenergizing the TR1 coil, opening the TR1 contact, deenergizing the R1 and TR2 coils. The R2 (N/O) contact closes, maintaining the R2 coil.

Compare the circuits in Figures 10-10 and 10-12. It will be noted that the circuits are compatible.

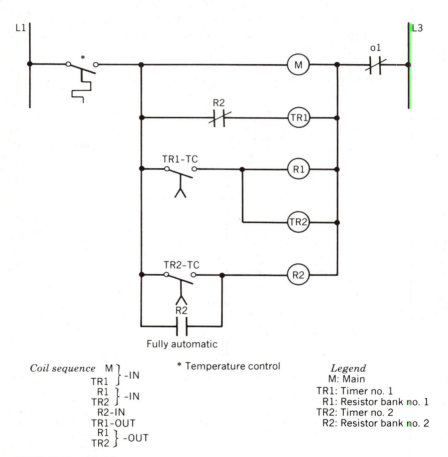

Coil sequence   M ⎫ -IN
         TR1 ⎭
         R1 ⎫ -IN
         TR2 ⎭
         R2-IN
         TR1-OUT
         R1 ⎫ -OUT
         TR2 ⎭

* Temperature control

Legend
M: Main
TR1: Timer no. 1
R1: Resistor bank no. 1
TR2: Timer no. 2
R2: Resistor bank no. 2

FIGURE   10-12

Three-Stage Secondary Resistance Starting, Nonreversing. Control Circuit. NVR Circuit (example two).

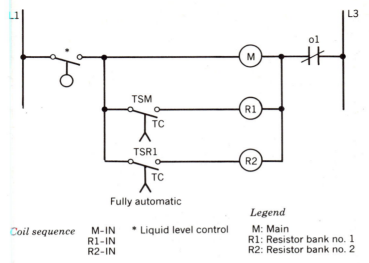

L1

L3

o1

M

TSM

TC

R1

TSR1

TC

R2

Fully automatic

Legend

Coil sequence   M-IN   * Liquid level control   M: Main
   R1-IN   R1: Resistor bank no. 1
   R2-IN   R2: Resistor bank no. 2

FIGURE 10-13

Three-Stage Secondary Resistance Starting, Nonreversing. Control Circuit. NVR Circuit (example three).

Control Circuit Operation, Figure 10-13

When the liquid level control closes, a circuit is completed for the M coil, releasing the TSM–TC contact. TSM–TC contact closes after a preset time delay, completing a circuit for the R1 coil, releasing contact TSR1–TC. Following a preset time delay, the contact TSR1 closes, completing a circuit for the R2 coil.

Compare the circuits in Figures 10–10 and 10-13. It will be noted that the circuits are compatible.

Control Circuit Operation, Figure 10-14

Pressing the start button completes a circuit for the M coil, which closes the maintaining contact M, and triggers (releases) the TSM contact. The TSM contact closes following the preset time delay, energizing the 1A coil, which releases the contact TS1A. The TS1A contact closes following the second preset time delay, energizing the 2A coil. The 2A (N/O) contact closes, maintaining the 2A coil, and the 2A (N/C) contact opens, deenergizing the 1A coil.

The circuits in Figures 10-11, 10-12, 10-13, and 10-14 could

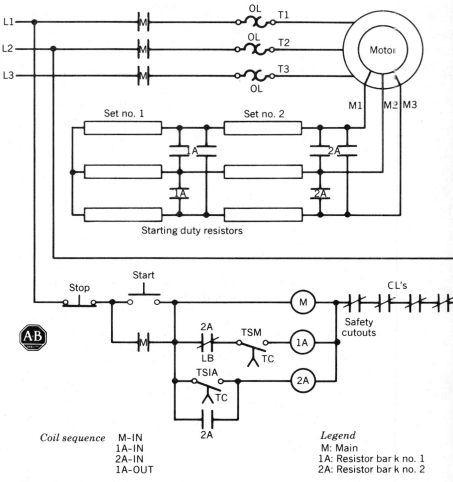

FIGURE 10-14

Three-Stage Secondary Resistance Starting, Nonreversing. Power and Control Circuits. NVP Circuit (example four).

be circuits designed by different motor control manufacturers. Learn to read, understand, and troubleshoot a given circuit without developing a personal preference for a particular one. Prepare yourself to be exposed to a variety of circuits that will perform the same operation, but differ in design and appearance.

REVERSING THREE-STAGE STARTING

Power Circuit Operation, Figure 10-15

When either the F or R power contacts close as illustrated in Figure 10-15, line voltage is applied to the stator winding. Both resistor banks are connected to the slip-rings. Starting current is kept to a lower level than it would be if line voltage were applied to the stator winding, and the slip-rings were shorted together. Following a preset time delay, the power contacts R1 will close, causing the motor speed to increase. The R2 power contacts will close following a second time delay, allowing the motor to run at rated speed.

To change the direction of rotation of a wound rotor induction motor, interchange any two phase conductors. The reversing magnetic motor starter will interchange the phase conductors. The forward and reverse contactors are mechanically interlocked. In addition to the mechanical interlock, electrical interlocks have been included in the control circuit.

Various control circuits for three-stage, reversible, secondary resistance starting are shown in Figures 10-16 and 10-17. Either of the circuits will cause the power circuit to perform as discussed in the power circuit operation for Figure 10-15.

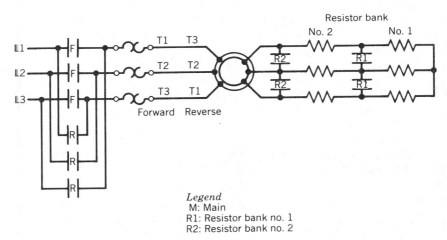

Legend
  M: Main
  R1: Resistor bank no. 1
  R2: Resistor bank no. 2

FIGURE   10-15

Three-Stage Secondary Resistance Starting, Reversible. Power Circuit.

Control Circuit Operation, Figure 10-16

Pressing the reverse pushbutton completes a circuit for the R coil, which opens the R(N/C) electrical interlock. Both R (N/O) contacts close. One R (N/O) contact maintains the circuit, and the second R (N/O) contact completes a circuit for the TR1 coil, releasing the TR1–TC contact.

Following a preset time delay, the TR1–TC contact closes, energizing the R1 and TR2 coils, releasing the TR2–TC contact. When the second preset time delay has expired, the TR2–TC contact closes, energizing the R2 coil.

The circuit in Figure 10-16 is designed in such a way that the stop pushbutton must be pressed before the opposite rotation may be selected.

Compare the circuits in Figures 10-15 and 10-16. It will be noted that the circuits are compatible.

Control Circuit Operation, Figure 10-17

The major difference between the circuits in Figures 10-16 and 10-17 is the electrical interlock in the remote pushbutton station. With this circuit, the opposite rotation may be selected without pressing the stop pushbutton.

Crane circuits employ electrical interlocks in the remote pushbutton stations. The instant the opposite rotation is selected, the starting resistors are placed into the rotor circuit, which will restrict the high currents in the line conductors. It should also be noted that the circuit has two timer coils and contacts, in addition to a triggered timing unit.

Compare the circuits in Figures 10-15 and 10-17. It will be noted that the circuits are compatible.

TWO-STAGE, TWO-SPEED, NONREVERSING REGULATOR

Figure 10-18 is an example of how motor control can be used to perform required functions and operations. There seems to be no end to what can be accomplished. The example presented takes a two-stage, secondary resistance starter, and by adding a control relay and altering the placements of contacts, a two-speed regulator has been developed.

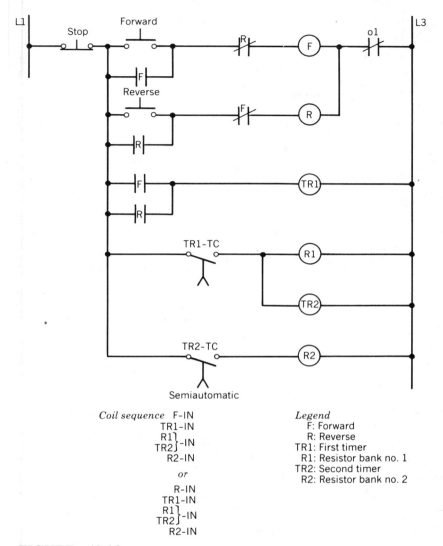

Coil sequence   F-IN
TR1-IN
R1⎫
TR2⎭ -IN
R2-IN

or

R-IN
TR1-IN
R1⎫
TR2⎭ -IN
R2-IN

Legend
F: Forward
R: Reverse
TR1: First timer
R1: Resistor bank no. 1
TR2: Second timer
R2: Resistor bank no. 2

FIGURE   10-16

Three-Stage Secondary Resistance Starting, Reversible. Control Circuit. NVP Circuit (example one).

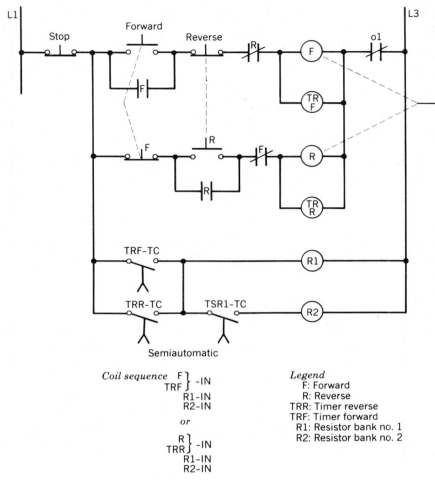

FIGURE 10-17

Three-Stage Secondary Resistance Starting, Reversible. Control Circuit. NVP Circuit (example two).

Circuit Operation

In operation, by pressing the high-speed pushbutton, a circuit is completed for the CR coil, which closes two CR contacts. One CR contact maintains the M and TR coils. The second CR contact maintains the circuit for the CR coil.

When the M coil is energized, it closes the M power contacts, causing the motor to start. Following the preset time

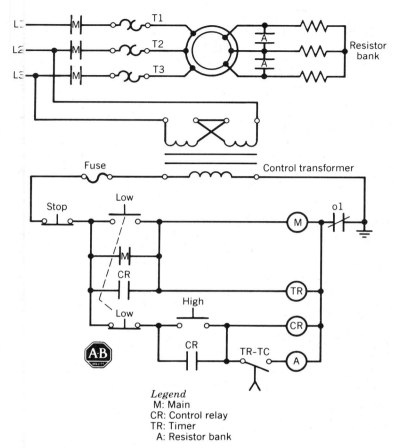

FIGURE 10-18

Secondary Resistance Starting. Two-Stage, Nonreversing, High-, Low-Speed Regulator. Power and Control Circuits.

delay, the TR–TC contact closes, energizing the A coil, closing the A resistor contacts, bypassing the resistors. The motor is now running at high speed.

Pressing the low pushbutton would deenergize the CR and A coil, opening the CR contacts and the A resistor contacts, placing the resistors back into the rotor circuit, causing the motor to run at low speed.

With the motor stopped, and by pressing the low pushbutton, a circuit is completed for the M and TR coils, closing the M power contacts, causing the motor to start and remain in the low speed. Following the preset time delay, the TR–TC contact closes, but the CR contact is open, making it impossible to complete the circuit for the A coil. The motor remains at low speed until the high pushbutton is pressed, at which time the motor will go directly into high speed.

HIGH-LIMIT (HIGH TEMPERATURE) PROTECTION

Resistors designed for starting the wound rotor induction motor could overheat if left connected in the rotor circuit for an extended period of time. A defective contact or a broken conductor could cause the resistors to remain in the circuit.

A high-limit is often mounted adjacent to the resistors. When the resistors overheat, the high-limit, which is connected in series with the overload contact, opens, thereby deenergizing the circuit. The high-limit is manually reset.

POWER CIRCUIT CALCULATION

Single Motor Installation

To design the power circuit for a single motor installation as shown in Figure 10-19, the following calculations must be understood:

1. Overcurrent protection.
2. Overload protection.
3. Conductor sizes
    A. Primary circuit conductors.
    B. Secondary circuit conductors.

Overcurrent Protection
The branch-circuit overcurrent device is to be rated at not more than the percentage of the FLA rating of the motor shown in Table 10-2.

Overload Protection
When selecting the correct overload device, obtain the full-load current from the motor nameplate. Apply the FLA to the overload chart that is located on the cover or door of the motor starter and select the proper overload device by catalogue number. Keep in mind the service factor of the motor.

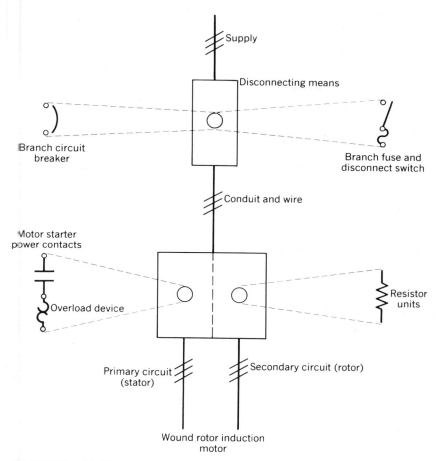

FIGURE 10-19

Single-Line Drawing of the Power Circuit for a Wound Rotor Motor Installation.

If the service factor (SF) is 1.0, the maximum value of the overload device (in amperes) will be 115% of the FLA. With a service factor of 1.15, the maximum value of the overload device will be 125% of the FLA.

Conductor Sizes

Refer to Figures 10-19 and 10-1. It should be noted that with a secondary resistance starting installation, two sets of three conductors are required from the motor to the motor starter and resistor bank.

TABLE 10-2

Canadian Electrical Code, Part 1

(TABLE 29)

(See Rules 28-200, 28-204, 28-208, and 28-210*)

RATING OR SETTING OF OVERCURRENT DEVICES FOR THE PROTECTION OF MOTOR BRANCH CIRCUITS

Type of Motor	Percent of Full-Load Current		
	Maximum Fuse Rating		Maximum Setting Time-Limit Type Circuit Breaker
	Time-Delay* "D" Fuses	Non-Time Delay	
Alternating Current			
Single-Phase, all types	175	300	250
Squirrel-Cage and Synchronous:			
Full-Voltage, Resistor and Reactor Starting	175	300	250
Autotransformer Starting:			
Not more than 30 A	175	250	200
More than 30 A	175	200	200
Wound Rotor	150	150	150
Direct Current	150	150	150

* (Except as permitted in Table 26 where 15-A overcurrent protection for motor branch-circuit conductors exceeds the values specified in the above table.)

** *Time-delay "D" fuses are those referred to in Rule 14-200.[a]*

NOTES: (1) *The ratings of fuses for the protection of motor branch circuits as given in Table 26 are based on fuse ratings appearing in the table above, which also specifies the maximum settings of circuit breakers for the protection of motor branch circuits.*

(2) *Synchronous motors of the low-torque low-speed type (usually 450 rpm, or lower) such as are used to drive reciprocating compressors, pumps, etc., and which start up unloaded, do not require a fuse rating or circuit-breaker setting in excess of 200% of full-load current.*

(3) *For the use of instantaneous trip (magnetic only) circuit interrupters in motor branch circuits see Rule 28-210.[a]*

[a] The (Table 29) and the Rule numbers mentioned refer to the Canadian Electrical Code. Similar comments will be listed in the National Electrical Code. With permission of Canadian Standards Association, this table is reproduced from CSA Standard C22.1-1982, Canadian Electrical Code, Part 1 (14th edition), which is copyrighted by CSA and copies may be purchased from CSA, 178 Rexdale Blvd., Rexdale, Ontario M9W 1R3.

Primary Conductors Refer to Figure 10-19. The minimum allowable ampacity of the power circuit conductors will be 125% of the FLA of the motor.

If, for example, the nameplate indicates 180 A, the minimum allowable ampacity of the primary conductors to the stator would be 180 A × 125% = 225 A. For standardization, 90 C copper conductors will be used. Therefore, the code will require a minimum of 4/0 conductors.

Secondary Conductors Many variables must be considered before calculating the minimum ampacity of the secondary conductors. Obtain the full-load secondary current of the motor from the engineering data for the installation, Table 10-1, or from the motor nameplate (if indicated).

The resistor "duty" classification must be determined. Apply the correct percentage and select the correct conductors. Keep in mind the requirements for a higher temperature insulation on conductors connected to resistors. The percentages and required insulation will be clearly listed in the electrical code.

REVIEW EXERCISES

10-1   Which part of a wound rotor induction motor is the primary circuit?

10-2   Which part of a wound rotor induction motor is the secondary circuit?

10-3   How are the overload devices selected?

10-4   How is the minimum ampacity of the power circuit conductors for the primary circuit calculated?

10-5   What does motor service factor mean?

10-6   Explain why full voltage is applied to the primary circuit when starting the motor, rather than reduced voltage.

10-7   What is the purpose of connecting resistance units in the secondary circuit during start-up?

10-8   How is the direction of rotation of a wound rotor induction motor reversed?

10-9 What is the purpose of installing a high-limit safety control adjacent to the resistors?

10-10 What is meant by three-stage starting?

10-11 Is secondary resistance starting open or closed transition?

10-12 Is this method of starting reduced voltage or reduced current starting?

10-13 How many conductors are required from the motor starter to the motor for the primary and secondary circuits?

PROBLEMS

The problems are divided into the following three categories:

Part A The application of the electrical code.

Part B Troubleshooting of motor control circuits.

Part C Multiple-choice questions on the combined topics.

Answers and solutions for the odd-numbered problems will be located at the back of this text. Answers and solutions for the even-numbered problems will be contained in the Instructor's Manual.

PART A
ELECTRICAL CODE

Instructions: When answering the following problems, keep in mind:

1. The starting method will be secondary resistance.

2. The overload device will be the maximum value in amperes.

3. The conductors will be R90 X-Link CU.

Refer to Figure 10-1.

1. Calculate
   A. The minimum conductor ampacity for the power circuit and

   _____

   _____

   B. Select the minimum size conductor for the primary circuit when the motor FLA is 75 A.

   _____

   _____

2. Calculate the maximum value in amperes of the overload device if the FLA of the motor is 71 A, with a service factor of 1.15.

   _____

   _____

3. Select the maximum allowable rating for time-delay fuses to be installed for a wound rotor induction motor if the FLA of the motor is 70 A.

   _____

4. The FLA of a wound rotor induction motor is 52 A. Calculate the value of the overload device if the service factor is 1.0.

   _____

PART B
TROUBLESHOOTING

Instructions: For this exercise, a fault will be stated. Study the designated circuit to determine how the circuit will operate.

1. See Figure 10-8. The TR–TC contact is defective and will not close. Press the start button. How will this fault affect the power circuit?

   _____

   _____

   _____

   _____

2. See Figure 10-8. The N/C R contact is defective and remains in the open position. Press the start button. Explain how the control circuit will operate.

_____

_____

_____

_____

_____

3. See Figure 10-11. The TR1–TC and TR2–TC contacts are stuck in the closed position. Press the start button. Explain what will happen to the primary winding current.

_____

_____

_____

4. See Figure 10-12. The thermostat has closed and the starting sequence has been completed. The motor is running with the rotor short-circuited. The following operations are conducted separately:

   A. Bridge the TR1–TC contact. Discuss the results.

   _____

   _____

   _____

   B. Bridge the TR2–TC contact. Discuss the results.

   _____

   _____

   _____

   C. Bridge the N/C R2 contact. Discuss the results.

   _____

   _____

   _____

   _____

   _____

D. Short out the R1 coil. Discuss the results.

_____

_____

E. Open-circuit the N/O R2 contact. Discuss the results.

_____

_____

_____

_____

F. Short out the R2 coil. Discuss the results.

_____

_____

G. Trip the overload contact. Discuss the results.

_____

_____

_____

PART C
MULTIPLE-CHOICE QUESTIONS

Instructions:

A. The electrical code book may be used to assist you in solving the following problems.

B. Select answers from a, b, c, or d.

C. Answer all questions, based on the drawing Figure 10-20.

1. The method of starting is:
   A. Full voltage.
   B. Primary resistance.
   C. Wye-delta.
   D. None of the above.

2. In the forward operation, the starting coil sequence would be:

A. R–IN
TR–IN
S1–IN
S2–IN

B. F–IN
TR–IN
S1–IN
S2–IN

C. F–IN
TR–IN
S1–IN
S2–IN
TR–OUT

D. F⎫ –IN
TR⎭
S1–IN
S2–IN

3. The starting method is:
A. Three-stage.
B. Two-stage.
C. Two-step.
D. Four-stage.

4. The overload devices are installed to protect the
A. Slip-rings.
B. Conductors.
C. Motor frame.
D. Stator windings.

5. The starting method is:
A. Open transition.
B. Taken off the line.
C. Overcurrent protection.
D. Closed transition.

6. The overload devices are selected by:
A. FLA × 150%.
B. FLA × 125%.
C. FLA applied to the chart in the motor starter.
D. Looking in the electrical code book.

7. The maximum rating for the overcurrent device for this method of starting is:
A. 150% of FLA.
B. $\dfrac{FLA}{2} \times 200\%$.
C. FLA × 300%.
D. Not specified in code book.

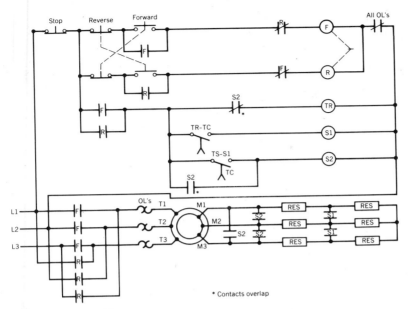

FIGURE 10-20

8. When the reverse button has been pressed and the motor is running on the first stage, which coils are not energized?

   A. F, TR, S1, and S2.
   B. TR, S1, and S2.
   C. S1 and S2.
   D. F, S1, and S2.

9. The motor is running in the forward rotation. An overload contact opens. Which coil is deenergized first?

   A. TR.
   B. S1.
   C. R.
   D. F.

10. The motor is running in the forward rotation with all resistance units bypassed. How many contacts in the control circuit are in the open position?

   A. 7.
   B. 8.
   C. 6.
   D. 5.

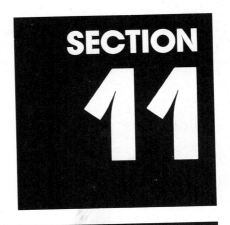

SECTION 11

DC MOTOR CONTROL

INTRODUCTION TO DC MOTOR CONTROL

Till this point, we have dealt primarily with the various aspects of AC motor control. This section will provide instruction on DC motor control. Topics such as timers and transition are common to both AC and DC motor control and were discussed in Section 5. Direct current motors have a definite advantage over alternating current motors; namely, speed regulation.

Although the use of DC motors is not as widespread as AC motors, the tradesperson should have a working knowledge of DC motor control. The three basic types of DC motors must be understood.

The rotating part of the motor is referred to as the armature, while the stationary part of the motor is referred to as the stator, which contains the series winding and the shunt winding.

SHUNT MOTOR

The shunt motor consists of a shunt field and the armature. Notice in Figure 11-1 that the shunt (field) winding and the armature are connected in parallel.

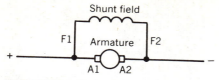

FIGURE 11-1

Connection for a Shunt Motor.

SERIES MOTOR

The series motor consists of a series (field) winding and the armature. Notice in Figure 11-2 that the series (field) winding and the armature are connected in series.

COMPOUND MOTOR

The compound motor is a combination of a shunt motor and a series motor. The series field and the armature are connected in series, with the shunt field connected in parallel with the series field and armature. See Figure 11-3.

Long Shunt Versus Short Shunt

Compound motors may be connected either long shunt or short shunt. Figure 11-3 illustrates the connection for long shunt, Figure 11-4 is the connection for short shunt.

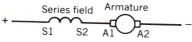

FIGURE 11-2

Connection for a Series Motor.

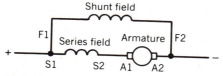

FIGURE 11-3

Connection for a Long Shunt
Compound Motor.

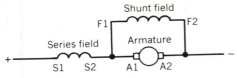

FIGURE 11-4

Connection for a Short Shunt Compound Motor.

CURRENT FLOW

Electron current flow is considered to be in the direction of negative (−) to positive (+), while conventional current flow is from positive (+) to negative (−). Throughout this section, electron flow has been used. If the reader chooses to use conventional flow, simply reverse the directional arrows to indicate conventional current flow.

TO REVERSE THE DIRECTION OF ROTATION OF A DC MOTOR

To reverse the direction of rotation of a shunt motor, change the direction of the current flow through the shunt field or armature, but not both. See Figure 11-5.

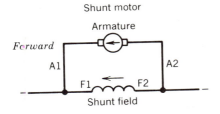

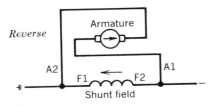

FIGURE 11-5

Connection for the Forward and the Reverse Rotation of a Shunt Motor.

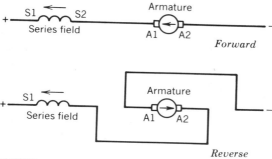

FIGURE   11-6

**Connection for the Forward and
the Reverse Rotation of a Series
Motor.**

To reverse the direction of rotation of a series motor,
change the direction of the current flow through the series
field or armature, but not both. See Figure 11-6.

To reverse the direction of rotation of a compound motor,
change the direction of the current flow through the armature.
See Figure 11-7.

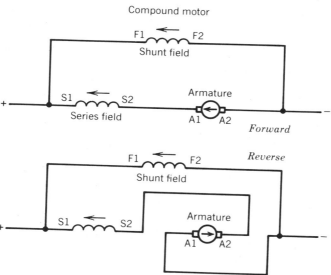

FIGURE   11-7

**Connection for the Forward and the Reverse Rotation of a Long
Shunt Compound Motor.**

ACROSS-THE-LINE MOTOR STARTING

This term means the same as "full voltage starting." When a DC motor is started on full line voltage, the starting current is generally very high.

DC motors up to 2 hp may be started on full line voltage or larger motors if the high in-rush current will not damage the motor. The motor manufacturer should be consulted.

POWER CIRCUIT

Before proceeding with the study of DC motor control, the power circuit must be fully understood. The electrical code requirements must be adhered to in all installations. We refer to such requirements as

1. Disconnecting means to isolate the motor and control equipment for safe maintenance of the driven machinery and electrical equipment.
2. The sizing of the overcurrent protection.
3. The selection and installation of the correct overload device.
4. The selection of the correct power circuit conductors.

Disconnecting Means

Disconnecting means for a motor installation may be in various forms. The electrical code divides the subject as follows:

1. Disconnecting means required.
2. Types of disconnecting means.
3. Rating of disconnecting means.
4. Location of disconnecting means.
5. Accessibility of disconnecting means.

Each of the above must be considered when designing an installation. The disconnecting means must be installed so that there will be no hazard to the tradesperson who is required to maintain the equipment. Think *safety*.

Overcurrent Protection

Overcurrent protection is installed in the power circuit to protect the conductors and is sized by the requirements of the electrical code. The percentage is shown in Table 11-1.

TABLE 11-1

Canadian Electrical Code, Part 1

(TABLE 29)

(See Rules 28-200, 28-204, 28-208, and 28-210*)

RATING OR SETTING OF OVERCURRENT DEVICES FOR THE PROTECTION OF MOTOR BRANCH CIRCUITS

Type of Motor	Percent of Full-Load Current		
	Maximum Fuse Rating		Maximum Setting Time-Limit Type Circuit Breaker
	Time-Delay* "D" Fuses	Non-Time Delay	
Alternating Current			
Single-Phase, all types	175	300	250
Squirrel-Cage and Synchronous:			
Full-Voltage, Resistor and Reactor Starting	175	300	250
Autotransformer Starting:			
Not more than 30 A	175	250	200
More than 30 A	175	200	200
Wound Rotor	150	150	150
Direct Current	150	150	150

* (Except as permitted in Table 26 where 15-A overcurrent protection for motor branch-circuit conductors exceeds the values specified in the above table.)

** *Time-delay "D" fuses are those referred to in Rule 14-200.[a]*

NOTES: (1) *The ratings of fuses for the protection of motor branch circuits as given in Table 26, are based on fuse ratings appearing in the table above, which also specifies the maximum settings of circuit breakers for the protection of motor branch circuits.*

(2) *Synchronous motors of the low-torque low-speed type (usually 450 rpm, or lower) such as are used to drive reciprocating compressors, pumps, etc., and which start up unloaded, do not require a fuse rating or circuit-breaker setting in excess of 200% of full-load current.*

(3) *For the use of instantaneous trip (magnetic only) circuit interrupters in motor branch circuits, see Rule 28-210.[a]*

[a] The (Table 29) and the Rule numbers mentioned refer to the Canadian Electrical Code. Similar comments will be listed in the National Electrical Code. With permission of Canadian Standards Association, this table is reproduced from CSA Standard C22.1-1982, Canadian Electrical Code, Part 1 (14th edition), which is copyrighted by CSA and copies may be purchased from CSA, 178 Rexdale Blvd., Rexdale, Ontario M9W 1R3.

An improper selection of an overcurrent device may result in damage to the equipment and possibly injury to personnel. The electrical code is very specific in regard to the selection of overcurrent devices.

Overload Protection

Overload protection is installed in the power circuit to protect the motor windings. The electrical code divides the subject as follows.

1. Overload protection required.
2. Overheating protection required.
3. Types of overload and overheating protection required.
4. Number and location of overload protective devices.
5. Shunting of overload protection during starting.
6. Automatically started motors.
7. Automatic restarting after overload.

SINGLE MOTOR INSTALLATION—POWER CIRCUIT

To determine conductor and conduit sizes, overcurrent and overload protection settings, solve the following.

EXAMPLE

See Figure 11-8.

Data:

Supply voltage	120 VDC
Motor horsepower	5 hp
Motor type	DC
Type of overcurrent device	Time-delay fuse
Starting method	Resistance
Wiring method	R90 X-Link CU in conduit

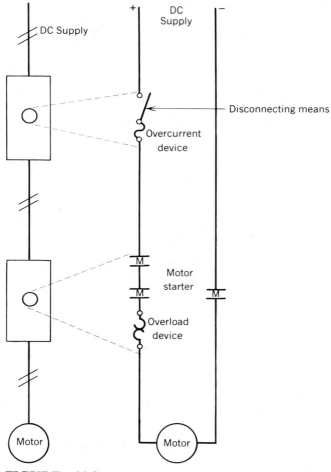

FIGURE   11-8

Single-Line Drawing of the Power Circuit for a Single DC Motor Installation. Full Voltage Starting.

Solution

Step 1   Obtain the estimated FLA from the electrical code book.

   Answer   40 A

Step 2   The minimum ampacity of the power circuit conductors will be 125% of FLA.

   Answer   125% of 40 A = 50 A

Step 3 Using the correct table in the electrical code, select the proper conductor size.

Answer No. 8 L R90 CU X-Link

Step 4 Using the correct table or tables in the electrical code, select the proper conduit size.

Answer 1/2 in. conduit (12 mm)

Step 5 To obtain the maximum time-delay fuse, consult the electrical code.

Answer 150% of 40 A = 60 A

Step 6 To obtain the correct overload heater, apply the FLA to the chart located on the cover of the motor starter.

Answer Select by catologue number. The maximum value of the heater in amperes will be 125% of FLA. 125% of 40 A = 50 A

The solutions have been entered in Table 11-2.

The complete installation may now be installed, using the information obtained from Steps 1 to 6.

MANUAL ACROSS-THE-LINE MOTOR STARTER

Figure 11-9 is a photo and wiring diagram of a standard manual "across-the-line" motor starter. The manual starter shown is unaffected by the loss of voltage and will remain in the on position if the supply voltage should fail. When the supply voltage is reestablished, the motor will restart automatically.

The term "no voltage release" generally refers to a control circuit involving an electromagnetic coil, but it also applies to this type of manual starter as well.

TABLE 11-2

Chart to Record Data for a Single Motor Installation, Figure 11-8.

Hp	FLA	Minimum Ampacity of Wire	Wire Size R90 X-Link	Conduit Size	Overload Device (value)	O/C Time-Delay Fuse
5	40 A	50 A	No. 8	$\frac{1}{2}$ in. (12 mm)	50 A	60 A

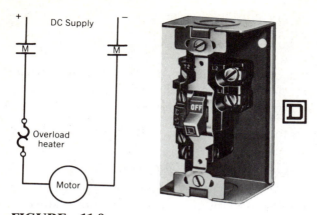

FIGURE 11-9

Manual Across-the-Line Motor Starter.

MAGNETIC ACROSS-THE-LINE STARTER

Unlike the manual motor starter, in which the power contacts are closed manually, the magnetic motor starter contacts are closed by energizing an electromagnetic coil. This allows for the fully automatic or semiautomatic control of machinery.

With a DC magnetic motor starter, it is important to realize that the breaking of the power circuit produces an arc, which will burn the power contacts if not extinguished quickly.

Blowout Coil or Equivalent

Some DC magnetic motor starters have a very low resistance coil connected above and in series with the main power contact. When the main power contact opens, the magnetic field in the blowout coil will collapse, drawing the arc upward, which will extinguish the arc. See Figure 11-10.

Other DC magnetic motor starters are equipped with two power contacts connected in series. See Figure 11-12 (positive line). However, other DC magnetic motor starters have two power contacts connected in series (positive line) and one power contact connected in the negative line, Figure 11-12. Still another DC magnetic motor starter employs a permanent magnet to draw the arc upward.

The method of extinguishing the arc is not important. It is important, however, to remember that some method has been devised to prevent damage to the power contacts. Recognize it and do nothing to defeat it.

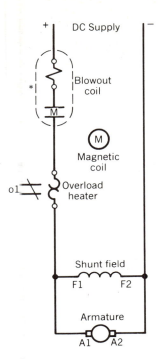

* Refer to the comments on blowout coil.

FIGURE 11-10

Magnetic Across-the-Line Motor Starter.

ELECTROMAGNETIC CONTROL

The contacts of a magnetic motor starter, contactor, or relay are closed by energizing an electromagnet. The electromagnet consists of a coil of wire on an iron core. See Figure 11-11.

When the current flows through the coil, the iron of the electromagnet becomes magnetized, attracting the armature.

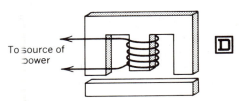

FIGURE 11-11

Electromagnet.

Interrupting the current flow through the coil causes the armature to drop out. The contacts are mechanically attached to the armature and will close and/or open as the armature moves.

NO VOLTAGE RELEASE
VERSUS NO VOLTAGE PROTECTION

When designing motor control circuits or troubleshooting equipment, two major terms must be fully understood.

A. No voltage release.

B. No voltage protection.

No Voltage Release

This term means that the motor will stop when there is a supply voltage failure, and the motor will restart automatically when the supply voltage is restored. The pilot device is uneffected by the loss of voltage, and its contacts will remain closed. The term "two-wire" control is applied to this type of circuit. Pilot devices such as float switches, limit switches, and temperature controls are single contacts, requiring two conductors.

Any piece of machinery that is automatically controlled may incorporate a no voltage release control circuit.

The following terms are generally accepted as having the same meaning:

A. Undervoltage release (UVR).

B. Low voltage release (LVR).

C. No voltage release (NVR).

No Voltage Protection

This term means that the motor will stop when there is a supply voltage failure and the motor will not restart automatically when the supply voltage is restored.

When automatic restarting is liable to create a hazard, the motor control device shall provide no voltage protection. Check this comment in the electrical code.

The following terms are generally accepted as having the same meaning:

A. Under voltage protection (UVP).

B. Low voltage protection (LVP).

C. No voltage protection (NVP).

Figures 11-12 and 11-13 illustrate a no voltage release circuit.

Circuit Operation, Figures 11-12 and 11-13

When the two-wire pilot device contact closes, a circuit is completed for the M coil, which closes all M power contacts. Full line voltage is connected to the motor terminals, and the motor operates. Opening the pilot device contact deenergizes the M coil, opening all M power contacts, stopping the motor. If the power supply should fail while the motor is operating, the M coil and the motor would be deenergized. On restoration of the supply voltage, the M coil would become ener-

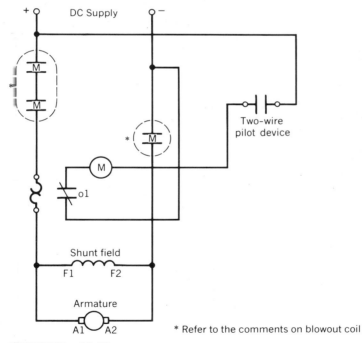

* Refer to the comments on blowout coil

FIGURE 11-12

**Wiring Diagram. Basic No
Voltage Release Circuit.
Power and Control.**

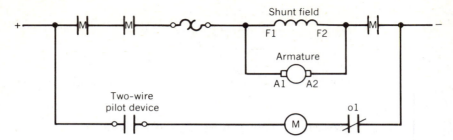

FIGURE 11-13

Schematic Diagram. Basic No Voltage Release Circuit. Power and Control.

gized, causing the motor to restart automatically. This is considered to be a fully automatic circuit.

Figures 11-14 and 11-15 illustrate a no voltage protection circuit.

Circuit Operation, Figures 11-14 and 11-15

Pressing the start button completes a circuit for the M coil, closing all M contacts. The three power contacts complete the circuit for the motor, and the M contact in the control circuit

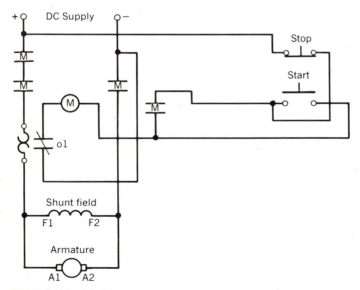

FIGURE 11-14

Wiring Diagram. Basic No Voltage Protection Circuit. Power and Control.

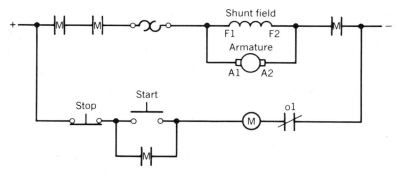

FIGURE 11-15

Schematic Diagram. Basic No Voltage Protection Circuit. Power and Control.

maintains the M coil when the momentary start button is released. If the supply voltage should fail, the M coil and the motor will be deenergized. On restoration of the supply voltage, the M coil would not be energized automatically. This is considered to be a semiautomatic circuit.

NO FIELD RELEASE

The circuit in Figure 11-14 has a built-in hazard. It is characteristic of a DC motor that if for any reason the shunt field should open-circuit and the motor is not loaded, the armature will pick up speed and could conceivably race to destruction.

A relay has been added to the control and power circuits (see Figures 11-16 and 11-17) that will monitor shunt field current. If the shunt field should open-circuit, the "field loss relay" (FLR) coil would cease to feel shunt field current and become deenergized, opening the FLR contact, deenergizing the M coil, disconnecting the motor from the line.

MOTOR SPEED

At the beginning of this section, a comment was made about DC motors and speed regulation. In order to control the speed of a DC motor, the shunt voltage or current must be varied, or the armature voltage or current. Figures 11-18, 11-19 and 11-20 illustrate the methods of controlling the speed of a DC motor.

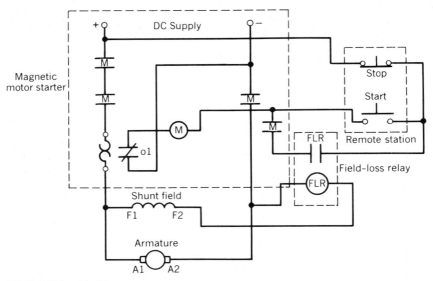

FIGURE 11-16

Wiring Diagram. Start–Stop Magnetic Motor Starter Circuit to Control a Shunt Motor. No Field Release Has Been Added to the Circuit.

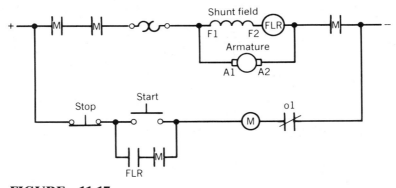

FIGURE 11-17

Schematic Diagram. Start–Stop Magnetic Motor Starter Circuit to Control a Shunt Motor. No Field Release Has Been Added to the Circuit.

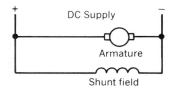

FIGURE 11-18
**DC Motor Connection for Base
Speed Operation.**

Base Speed

Base speed is when a DC motor is properly connected and
rated voltage is applied to the motor.

Above-Base Speed

Above-base speed is achieved by inserting resistance in the
shunt field.

Below-Base Speed

Below-base speed is achieved by inserting resistance in the
armature.

Speed Regulation

The circuit shown in Figure 11-21 provides above-base speed.
When the pilot device contact closes, the M coil is energized,
closing the M power contacts. If the field rheostat is bypassed, the motor will operate at base speed. Adding resistance to the shunt field causes a voltage drop across the field

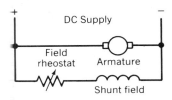

FIGURE 11-19
**DC Motor Connection for
Above-Base Speed Opera-
tion.**

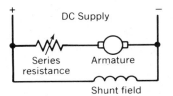

FIGURE 11-20
**DC Motor Connection for Be-
low-Base Speed Operation.**

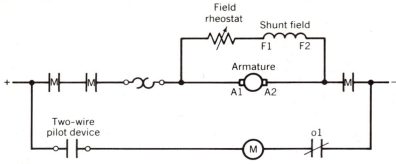

FIGURE 11-21

**Schematic Diagram. Magnetic Motor Starter Controlling a Shunt
Motor for Above-Base-Speed Operation.**

rheostat, reducing the voltage across the shunt field. The motor speed will increase.

DRUM SWITCHES

Drum switches may be used to control the direction of rotation of DC motors. The drum switches in Figures 11-22 and

FIGURE 11-22
Six-Point Drum Switch.

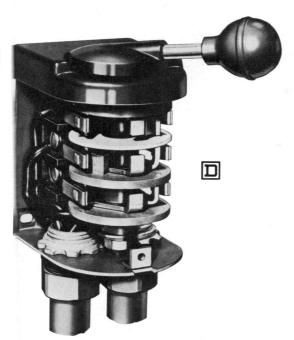

FIGURE 11-23

Six-Point Drum Switch (cover removed).

11-23 have a horsepower rating and three positions: reverse, off, and forward. Overload or overcurrent protection is not provided by a drum switch. The contacts are designed to be maintained or momentary and are easily converted in the field.

Drum switches are identified by their number of connection points. The three most common drum switches are

A. Six-point.
B. Eight-point.
C. Nine-point.

Drum Switch Contact Positions

The internal switching for each of the drum switches just described is illustrated in Figures 11-24 through 11-26.

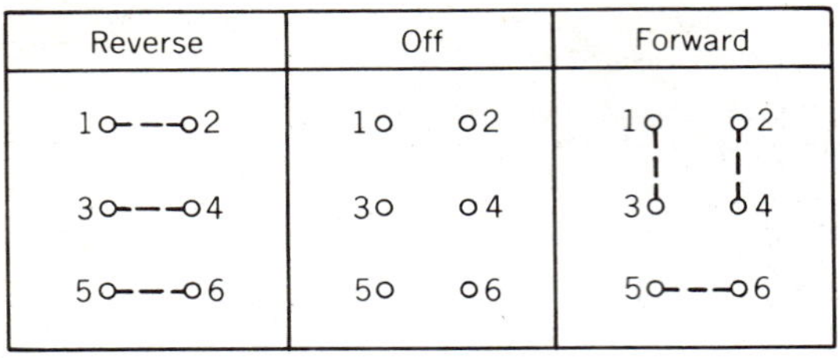

Reverse	Off	Forward
1o– –o2	1o o2	1o̧ ̧o2
3o– –o4	3o o4	3ȯ ȯ4
5o– –o6	5o o6	5o– –o6

FIGURE 11-24
Six-Point Drum Switch Contact
Positions.

Six-Point Drum Switch

Figures 11-27 through 11-29 are the circuit diagrams for DC motors controlled by six-point drum switches. Arrows have been used to indicate the direction of the current flow. It should be noted that the current flow remains unchanged through the shunt field and also the series field. Changing the direction of the current flow through the armature will reverse the direction of rotation.

Eight-Point Drum Switch

Figures 11-30 through 11-32 are the circuit diagrams for DC motors controlled by eight-point drum switches. Arrows have been used to indicate the direction of the current flow. It should be noted that the current flow remains unchanged through the shunt field and also the series field. Changing the

Reverse	Off	Forward
1o̧– –o2	1o o2	1o– –o̧2
3ȯ o̧4	3o o4	3o̧ ȯ4
5o– –ȯ6	5o o6	5ȯ– –o6
7o– –o8	7o o8	7o– –o8

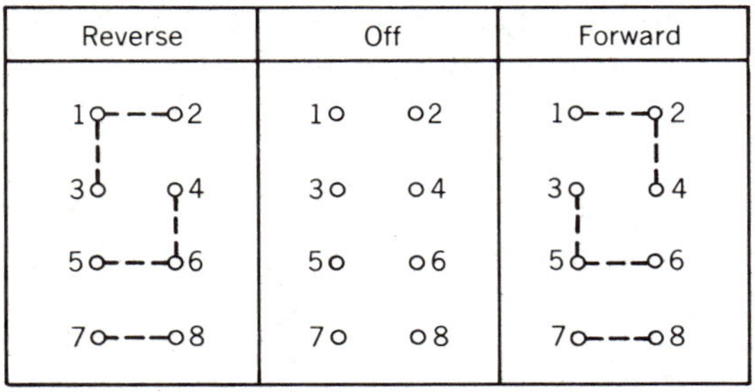

FIGURE 11-25
Eight-Point Drum Switch Contact Positions.

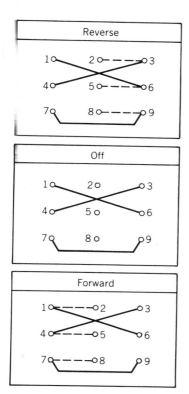

FIGURE 11-26

Nine-Point Drum Switch Contact Positions.

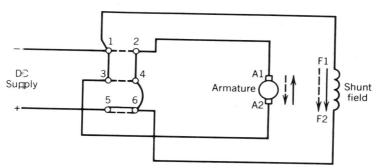

FIGURE 11-27

Shunt Motor Connected to a Six-Point Drum Switch.

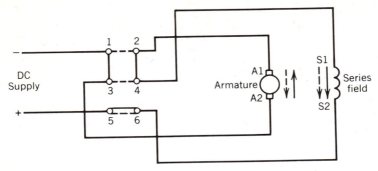

FIGURE 11-28

Series Motor Connected to a Six-Point Drum Switch.

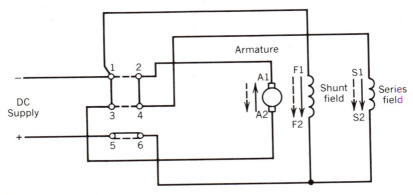

FIGURE 11-29

Compound Motor Connected to a Six-Point Drum Switch.

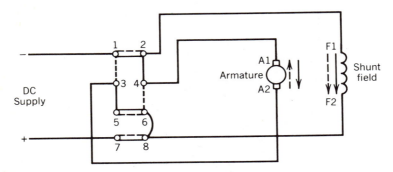

FIGURE 11-30

Shunt Motor Connected to an Eight-Point Drum Switch.

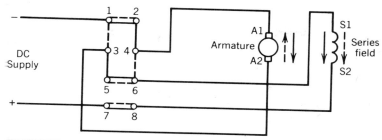

FIGURE 11-31

Series Motor Connected to an Eight-Point Drum Switch.

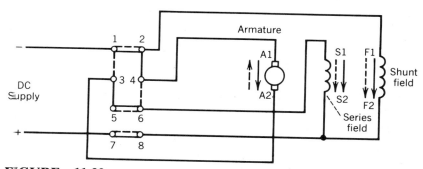

FIGURE 11-32

Compound Motor Connected to an Eight-Point Drum Switch.

direction of the current flow through the armature will reverse the direction of rotation.

Nine-Point Drum Switch

Figures 11-33 through 11-35 are the circuit diagrams for DC motors controlled by nine-point drum switches. Arrows have

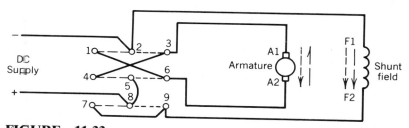

FIGURE 11-33

Shunt Motor Connected to a Nine-Point Drum Switch.

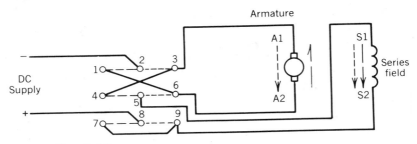

FIGURE 11-34

Series Motor Connected to a Nine-Point Drum Switch.

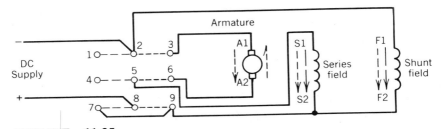

FIGURE 11-35

Compound Motor Connected to a Nine-Point Drum Switch.

been used to indicate the direction of the current flow. It should be noted that the current flow remains unchanged through the shunt field and also the series field. Changing the direction of the current flow through the armature will reverse the direction of rotation.

DOUBLE-POLE, DOUBLE-THROW REVERSING SWITCH

In addition to the drum switch for reversing a DC motor, a double-pole, double-throw (DPDT) reversing switch may be used for selecting the required rotation. The DPDT switch is often referred to as a field-reversing switch. Unlike the drum switches that may be used to start and stop the motor, this switch does not have an off position, so it selects the rotation only.

Figure 11-36 illustrates the contact positions for both the forward and reverse rotation. Figures 11-37 through 11-39 show the circuit diagrams for a shunt motor, a series motor, and a compound motor using a DPDT switch for reversing the rotation.

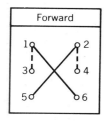
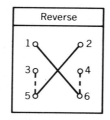

FIGURE 11-36

Contact Positions for a Double-Pole, Double-Throw Reversing Switch (with no off position).

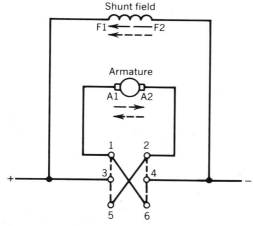

FIGURE 11-37

Shunt Motor Connected to a DPDT Reversing Switch.

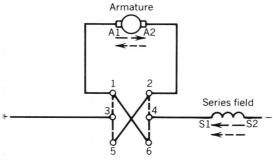

FIGURE 11-38

Series Motor Connected to a DPDT Reversing Switch.

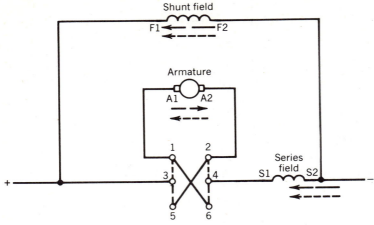

FIGURE 11-39

Compound Motor Connected to a DPDT Reversing Switch.

Arrows have been used to indicate the direction of the current flow. It should be noticed that the current flow remains unchanged through the shunt field and the series field. Changing the direction of the current flow through the armature will reverse the direction of rotation.

REVERSING MAGNETIC MOTOR STARTER

Drum switches provide a selection of rotation, but will not provide automatic control, remote control, or overload protection. Reversing magnetic motor starters will provide overload protection and various automatic and semiautomatic operations.

Figure 11-40 is a wiring diagram showing the power circuit for a reversing magnetic motor starter to operate a shunt motor. The arrows indicate the electron (current) flow in both the armature and the shunt field.

Constant Speed

Figure 11-41 is a schematic diagram illustrating a typical power and control circuit for the forward and reverse operation of the motor.

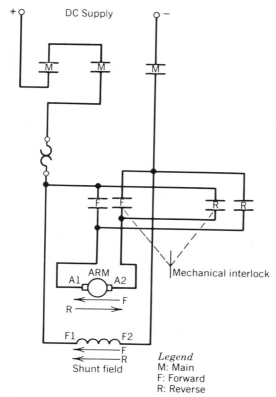

FIGURE 11-40

Reversing Magnetic Motor Starter to Reverse a Shunt Motor. Power Circuit.

Circuit Operation

Pressing the forward button completes a circuit for the F coil, closing all F contacts and opening the N/C F contact. One N/O F contact bypasses the forward button, maintaining the circuit for the F coil. Another N/O F contact has closed, energizing the M coil. The motor has been connected directly across the line.

The N/C F interlock prevents the R coil from being energized while the F coil is in operation. Pressing the stop button deenergizes both coils and the motor stops. Pressing the reverse button completes the circuit for the reverse rotation.

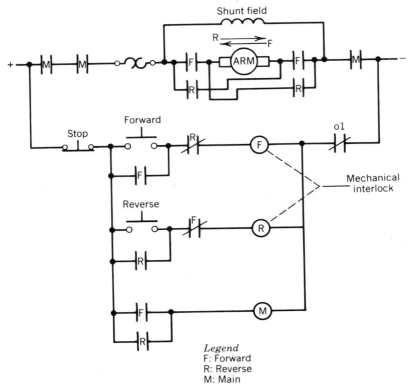

FIGURE 11-41

Schematic Diagram. Reversing a Shunt Motor with a Reversing Magnetic Motor Starter. Constant Speed. Power and Control.

Adjustable Speed Operation

When starting a motor that is to be operated at above-base speed, care must be taken to ensure that the speed regulator rheostat is in the bypassed position. Failing to bypass the rheostat will cause high current to flow in the armature, due to a weak shunt field.

Figure 11-42 depicts a circuit for a reversing magnetic motor starter to control a shunt motor, which is to operate at above-base speed. The N/C FR contact has bypassed the field rheostat and will not open until the preset time delay has expired, energizing the FR coil. The FR contact opens, allowing above-base speed operation. The remainder of the circuit operates the same as the circuit in Figure 11-41.

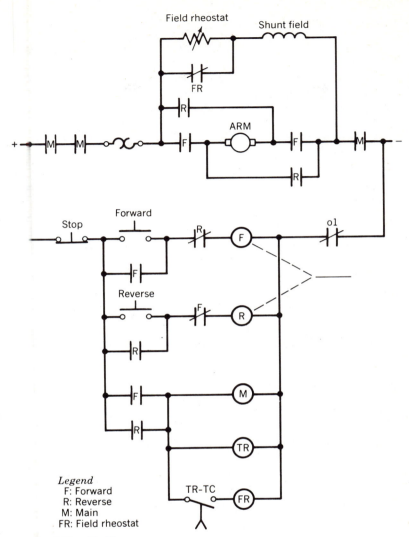

FIGURE 11-42

Schematic Diagram. Reversing a Shunt Motor with a Reversing Magnetic Motor Starter. Adjustable Speed. Power and Control.

ANTI-PLUGGING

Plugging as related to DC motor control is defined as the attempt to reverse the direction of rotation of a DC motor, without allowing the motor to approach zero speed. If this were permitted, the extremely high current would cause damage to the armature winding, commutator, and/or brushes.

To prevent plugging, a control called an "anti-plugging re-lay" is installed as shown in Figure 11-43. The anti-plugging relay consists of an electromagnetic coil and four contacts: two normally open and two normally closed.

Refer to the circuit diagram in Figure 11-43. The two N/O AP contacts are connected in series to control the shunt field. Note the placement of the N/C AP contacts.

Circuit Operation

The operation of the circuit in Figure 11-43 is the same as the circuit shown in Figure 11-40. The only difference is the operation of the anti-plugging (AP) relay.

Pressing the stop button deenergizes either the F or R coil

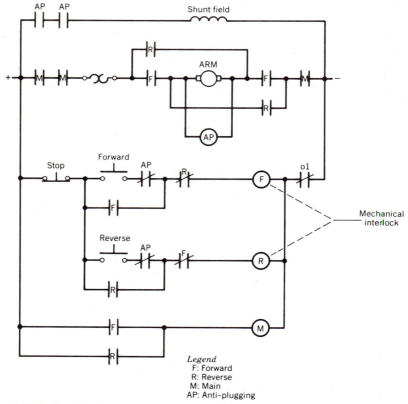

FIGURE 11-43

Anti-plugging. Power and Control.

and the M coil. The N/O AP contacts remain closed until the AP coil is deenergized.

With the shunt field still energized and the momentum of the rotating machinery, the armature continues to rotate. The motor while slowing down to zero speed is actually a generator, which keeps the AP coil energized.

On reaching zero speed, the motor is no longer a generator, deenergizing the AP coil, opening the circuit for the shunt field, reclosing the AP contacts in the control circuit, ready for the opposite rotation to be selected.

REDUCED VOLTAGE STARTING

When DC motors are started on full line voltage, the starting current in the armature is very high. Counterelectromotive force (CEMF) is generated by the DC motor, which will cause the current to drop, but until the CEMF is produced, the current is quite high.

To restrict the starting current of a DC motor during the start-up period, reduced voltage is applied to the armature. A variety of starting methods are available to reduce the starting voltage (current). The following circuits will demonstrate the various methods.

THREE-POINT MANUAL RESISTANCE STARTER

The three-point starter is often referred to as a faceplate starter. Three connection points are available; hence, the name "three-point starter." Examine Figure 11-44.

The purpose of the starter is to restrict starting current and also to provide no field release. The two main parts of the starter are a tapped resistance unit and a holding coil.

Circuit Operation

Moving the starting handle to the first contact point connects the full resistance in series with the armature. This resistance unit provides reduced voltage to the armature. A second circuit is completed for the shunt field through the holding coil (HC).

As the armature rotates, CEMF is generated, keeping the armature current at a tolerable level. The starting handle is

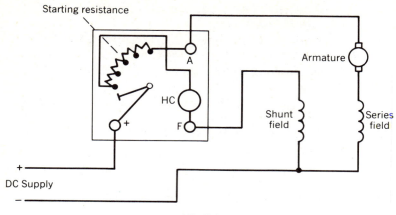

HC: Holding coil

FIGURE   11-44

Three-Point Manual Resistance Motor Starter Controlling a Compound Motor.

slowly moved across the starting contacts to the final resistance tap.

The starting handle, which is spring-loaded to the off position, is held at the final point by an electromagnet. It should be noted at this point that the resistance is now connected in the shunt field circuit.

The shunt field, which has a high ohmic resistance, will not be affected by the much lower starting resistance used in the armature circuit.

If the supply voltage should fail, the holding coil HC would be deenergized, releasing the handle to the off position. On resumption of the supply voltage, the motor will not start. The three-point starter provides no voltage protection.

If the shunt field should open-circuit, the holding coil would be deenergized, causing the handle to return to the off position. The three-point starter provides no field release.

A three-point starter is not designed to be used when controlling a motor above base speed. If a resistance were inserted in the shunt field to reduce shunt current, the holding coil could be weakened to a point at which the starting handle is released.

Overload and overcurrent protection are not provided by this starter. Additional equipment would be required to obtain such protection.

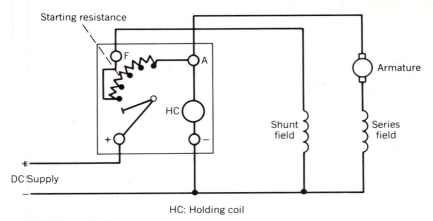

HC: Holding coil

FIGURE 11-45

Four-Point Manual Resistance Motor Starter Controlling a Compound Motor.

FOUR-POINT
MANUAL RESISTANCE STARTER

The four-point starter is similar to the three-point starter. Examine Figure 11-45. The similarity is that it will also provide no voltage protection and restrict starting current.

Four-point starters have four connection points, as compared to three for the three-point starter. Above-base speed is permissable with a four-point starter, as the holding coil is connected across the line and is not affected by the shunt current. If the shunt field should open-circuit, the motor would race to a dangerous speed.

TWO-STAGE
REDUCED VOLTAGE STARTER

Constant Speed

Figure 11-46 illustrates a typical two-stage resistance starter, referred to as a definite time-delay starter.

Power Circuit Operation

When the M power contacts close, full line voltage is applied to the shunt field, while the resistor is connected in series with

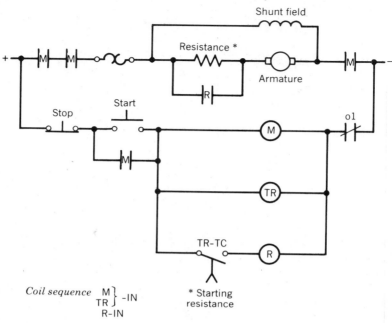

FIGURE 11-46

Two-Stage Reduced Voltage Starting. Constant Speed.

the armature. Following a preset time delay, the R contact closes, bypassing the resistor, allowing the motor to operate at base speed.

Control Circuit Operation

Pressing the start button energizes the M and TR coils, closing the M maintaining contact. Following a preset time delay, the TR–TC contact closes, energizing the R coil. The starting method is closed transition.

FOUR-STAGE REDUCED VOLTAGE STARTER

Adjustable Speed

Figure 11-47 shows a typical circuit, which does not depend on armature current to change from one stage to another. The

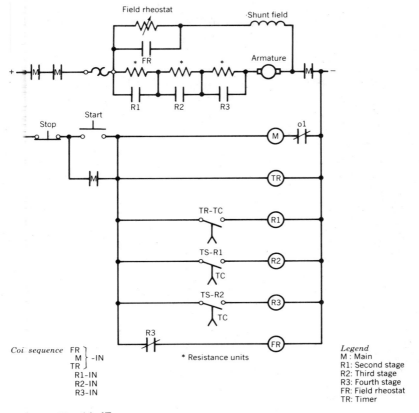

FIGURE 11-47

Four-Stage Reduced Voltage Starting. Adjustable Speed.

timer and timing units are preset and will operate regardless of the armature current or CEMF.

The motor installation provides variable above-base speed control. Care must be taken to ensure a strong shunt field; therefore, the field rheostat should be bypassed during the start-up period.

Power Circuit Operation

When the M contacts close, the shunt field is supplied with line voltage, and the resistors and armature are connected in series, across line voltage.

Following the first time delay, the R1 contact closes, bypassing a section of the resistance. Following the second time delay, the R2 contact closes, and following the third delay, the R3 contact closes. The resistors are all bypassed and the motor is running at base speed.

During the starting period, the FR contact remains closed. When the FR contact opens, the motor may be operated at above-base speed.

Control Circuit Operation

Pressing the start button energizes the M, TR, and FR coils. The M-maintaining contact maintains the circuit and the FR contact bypasses the field rheostat in the power circuit. Following a preset time delay, the TR–TC contact closes, energizing the R1 coil. After the second time delay, the TS-R1 contact closes, energizing the R2 coil. Following a third time delay, the TS-R2 contact closes, energizing the R3 coil, opening the N/C R3 contact, deenergizing the FR coil, opening the FR contact in the power circuit. The motor may now be operated at above-base speed. The starting method is closed transition.

DASHPOT ACCLERATION RESISTANCE STARTER

Constant Speed

Figure 11-48 illustrates a power circuit comparable to that in Figure 11-47. The difference is the method of bypassing the resistance in steps. Figure 11-47 employed contactors and timers; Figure 11-48 incorporates a dashpot timer.

Energizing the dashpot coil causes the rod to rise, moving the power contacts closer to the resistance. Contacts close in sequence: one, two, and three. Each contact bypasses a section of the resistance.

The oil in the dashpot retards the closing of the resistor contacts. An adjustable port allows the time to be shortened or extended.

COUNTERELECTROMOTIVE FORCE (CEMF) STARTER

The CEMF starter may be two-stage as shown in Figure 11-49 or multistage by using additional CEMF coils and contacts.

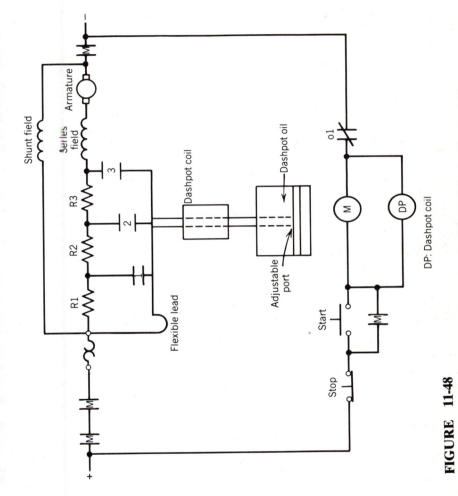

FIGURE 11-48

Dashpot Acceleration Resistance Starting Controlling a Compound Motor.

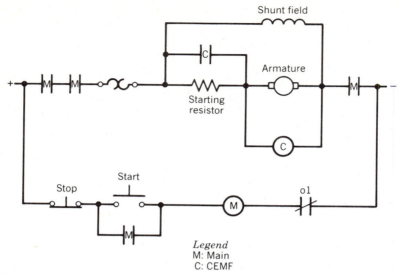

FIGURE 11-49

Counterelectromotive Force (CEMF) Reduced Voltage Starter.

Circuit Operation

Pressing the start button energizes the M coil, closing all M contacts. When the M power contacts close, the shunt field is connected directly across line voltage, and the resistor and the armature are connected in series, across line voltage.

The CEMF coil is connected across the armature. As the armature rotates, the CEMF increases. When the CEMF builds up, the C coil becomes a strong electromagnet, closing the C contact across the resistor, allowing the motor to operate at base speed.

This method of starting is not a definite time starter, as it depends on armature CEMF. The method of starting is closed transition.

TWO-COIL, LOCK-OUT RELAY, CURRENT-LIMIT STARTER

Constant Speed

This method of starting is reduced voltage starting, but it is dependent on armature current to operate the lock-out relay.

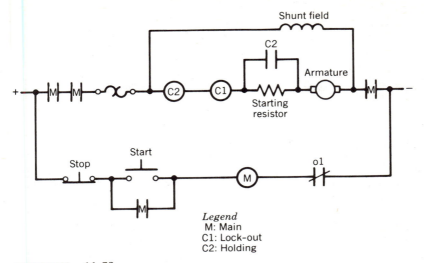

FIGURE 11-50

Two-Coil, Lock-Out Relay, Current-Limit Starting.

Figure 11-50 is the schematic for both the power and control circuit. The control circuit is the basic start–stop circuit.

Power Circuit Operation

When the M power contacts close, the shunt field is connected across line voltage. The C1 and C2 coils, which are both low resistance coils, are connected in series with the starting resistor and the armature across line voltage.

The holding coil has an iron core, which is easily saturated, while the lock-out coil has an air-gap that does not saturate. When the current is high, the lock-out magnetic field is stronger than the holding coil's magnetic field. This keeps the C2 contact open. As the CEMF builds up, the armature current drops to a point at which the holding coil's magnetic field is stronger, closing the C2 contact, bypassing the starting resistance.

This permits the motor to operate at base speed. The starting method is closed transition.

DYNAMIC BRAKING

Dynamic braking is a method of braking that uses the DC motor as a generator during the braking period. To understand

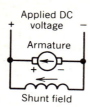

FIGURE 11-51

Shunt Motor in Operation.

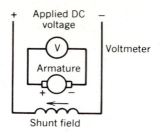

FIGURE 11-52

The Armature Disconnected and the Voltmeter Indicating Polarity.

dynamic braking, the student must first understand DC motor and generator current flow.

Figure 11-51 illustrates the connection for a shunt motor. Using electron flow (negative to positive), notice the direction of the current flow.

Figure 11-52 shows the armature removed from the applied voltage. The brushes remain (+) and (−), and the momentum of the armature has changed the motor into a generator. The voltmeter would indicate voltage.

Figure 11-53 depicts a variable external resistor connected across the armature. With the electron flow (negative to positive), the current would flow from the negative brush through the resistor to the positive brush. This would cause a current flow in the opposite direction, changing the polarity of the armature poles, stopping the motor quickly.

Figure 11-54 illustrates a typical power and control circuit using dynamic braking for stopping the motor quickly. The

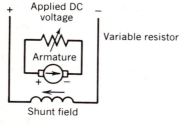

FIGURE 11-53

A Resistor Placed Across the Armature Causing Current to Flow.

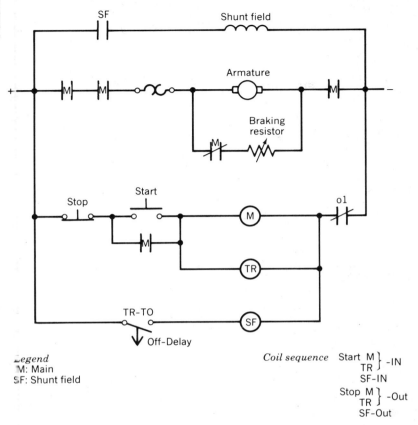

FIGURE 11-54

Dynamic Braking. Power and Control Circuit.

variable resistance allows the machine operator to control the armature current during the braking period, which controls the braking time.

Apply the comments presented for Figures 11-51, 11-52, and 11-53 when analyzing this circuit.

Circuit Operation

Pressing the start button energizes the M and TR coils. The M coil closes all M contacts. Three N/O M power contacts energize the armature circuit. The N/C M contact disconnects the circuit for the braking resistor, and a N/O M contact maintains the control circuit. At the same instant, the TR–TO contact closes, energizing the SF coil, closing the SF contact, connecting the shunt field across line voltage.

Pressing the stop button deenergizes the M coil, opening the N/O M power contacts, and also opening the M maintaining contact. At the same instant, the N/C M contact recloses.

While the armature is slowing down, the motor is operating as a generator. The external resistor provides a current path through the armature in the opposite direction, causing the motor to stop.

The shunt field remains energized until the TR–TO off time-delay contact opens, disconnecting the shunt field.

REVIEW EXERCISES

11-1 Name three types of DC motors.

11-2 How is the overload heater selected to protect a DC motor?

11-3 How many conductors are required from the motor starter to the motor when a shunt motor is started "across-the-line"?

11-4 Explain "no field release."

11-5 What is the function of the maintaining contact?

11-6 Explain how the overload heater and the overload contact operate.

11-7 Explain "base speed."

11-8 Explain "above-base speed."

11-9 Explain the function of the field rheostat and how the speed is controlled by the rheostat.

11-10 Compare conventional and electron current flow.

11-11 How is the direction of rotation of a shunt motor reversed?

11-12 Explain "dynamic braking."

11-13 Explain "anti-plugging."

11-14 Compare the three-point starter with the four-point starter.

11-15 Explain "dashpot acceleration resistance starting."

PROBLEMS

The problems are divided into the following three categories:

Part A The application of the electrical code.
Part B Troubleshooting of motor control circuits.
Part C Multiple-choice questions on the combined topics.

Answers and solutions for the odd-numbered problems will be located at the back of this text. Answers and solutions for the even-numbered problems will be contained in the Instructor's Manual.

PART A
ELECTRICAL CODE

Instructions: When answering the following problems, keep in mind:

1. The supply voltage is 120 V DC.
2. The starting method is resistance starting.
3. The overload device will be the calculated value in amperes.

Refer to Figure 11-8.

1. Calculate the minimum conductor ampacity for the power circuit if the motor FLA is 68 A.

2. Calculate the overload heater value if the motor FLA is 95 A.

3. Select the minimum conductor for the power circuit if the motor FLA is 78 A.

4. Select the maximum time-limit circuit breaker to be installed to protect the branch-circuit conductors if the motor FLA is 100 A.

   _____

   _____

5. Select the minimum size conduit from the disconnecting
   means to the motor starter for the power circuit conductors if the motor FLA is 40 A.

   _____

   _____

6. Select the maximum time-delay fuse to be installed to
   protect the power circuit conductors if the motor FLA is
   80 A.

   _____

   _____

PART B
TROUBLESHOOTING

Instructions: For this exercise, a fault will be stated. Study
the designated circuit to determine how the circuit will operate.

1. See Figure 11-17. The motor is running. If the shunt field
   should open-circuit, how will the circuit react?

   _____

   _____

   _____

2. See Figure 11-17. The FLR contact is defective and will
   not close. Press the start button. Discuss the results.

   _____

   _____

   _____

3. See Figure 11-21. The motor is running. If the brush at A1

should break, causing an open circuit for the armature, how will the motor react?

4. See Figure 11-41. The N/C R contact in the control circuit is defective and will not close. Press the reverse button. How will the motor operate?

5. See Figure 11-42. A serious malfunction in the motor starter has allowed the F and R power circuit contacts to close at the same time. Press the forward button. Discuss the circuit operation.

6. See Figure 11-47. The TR–TC contact is welded in the closed position. Press the start button. Explain the operation of the power circuit.

7. See Figure 11-49. The N/O M maintaining contact is defective and will not close. Press the start button momentarily. How will the circuit respond?

8. See Figure 11-54. The N/C M contact is defective and will not close. Press the start button. Discuss the circuit operation when the stop button is pressed.

PART C
MULTIPLE-CHOICE QUESTIONS

Instructions:

A. The electrical code book may be used to assist you in solving the following problems.

B. Select answers from a, b, c, or d.

C. Answer all questions based on the drawing in Figure 11-55.

1. The starting method is:
   - **A.** Dynamic braking.
   - **B.** Resistance.
   - **C.** Three-coil lockout.
   - **D.** Across-the-line.

2. The type of transition is:
   - **A.** Closed.
   - **B.** Open.
   - **C.** Combination of both a and b.
   - **D.** Overload.

3. The overload device protects the:
   - **A.** Conductors.
   - **B.** Resistors.
   - **C.** Motor.
   - **D.** Field rheostat.

4. The control circuit provides:
   - **A.** Low voltage release.
   - **B.** No field release.
   - **C.** Field loss protection.
   - **D.** No voltage protection.

5. The coil sequence is:
   - **A.**   M–IN
         TR–IN
         R1–IN
         R2–IN
         R3–IN
   - **B.** $\left.\begin{array}{c}\text{M}\\\text{TR}\end{array}\right\}$–IN
      $\left.\begin{array}{c}\text{R1}\\\text{R2}\end{array}\right\}$–IN
      R3 –IN

C. TR–IN
   M–IN
   R1–IN
   R2–IN
   R3–IN

D. $\left.\begin{array}{c} FR \\ M \\ TR \end{array}\right\}$–IN
   R1–IN
   R2–IN
   R3–IN

6. When the motor is running with all resistance units by-passed, how many coils are energized?
   A. 3.
   B. 5.
   C. 6.
   D. 4.

7. The motor is running and the overload contact opens. Which coil becomes deenergized first?
   A. M.
   B. TR.
   C. R1.
   D. None of the above.

8. The overload heater for this circuit is selected by:
   A. FLA to the overload chart.
   B. FLA to the electrical code book.
   C. FLA × 125%, to the overload chart.
   D. FLA × 15%, to the electrical code book.

9. The maximum size overcurrent device is determined by:
   A. FLA × 175%.
   B. FLA × 125%.
   C. FLA × 500%.
   D. FLA × 150%.

10. The TS–R1–TC contact is defective and will not close. If we press the start button,
    A. Two sets of resistance units will remain in the armature circuit.
    B. The FR coil will be deenergized.
    C. The overload contact will open.
    D. The motor will operate above base speed.

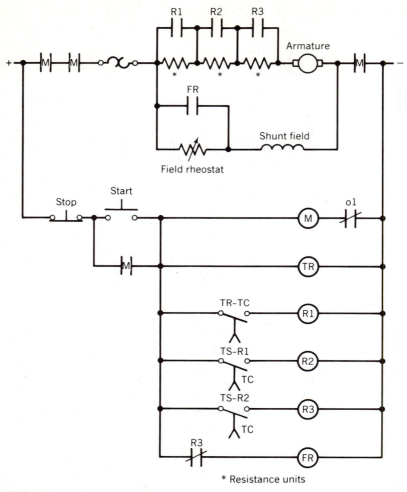

FIGURE 11-55

11. Increasing the resistance in the shunt field causes the motor
   A. To operate below base speed.
   B. To operate at base speed.
   C. To operate above base speed.
   D. To slow down.

12. Increasing the resistance in the armature causes the motor

   A.   To increase in speed.
   B.   To decrease in speed.
   C.   To run at base speed.
   D.   To run above base speed.

APPENDIX

SECTION 6, ANSWERS AND SOLUTIONS

PART A
ELECTRICAL CODE

1. **A.** $82 \text{ A} \times 125\% = 102.5 \text{ A}.$
    B.   No.3 R90 X-Link CU.

3. $62 \text{ A} \times 200\% = 124 \text{ A}.$
Therefore, select a 125-A circuit breaker.

PART B
TROUBLESHOOTING

1. Nothing would happen.
3. One (coil R).
5. The 1S coil would be energized, and possibly the 2S coil as well. The circuit would cease to operate.
7. The circuit would shut down.

PART C
MULTIPLE-CHOICE QUESTIONS

1. A.
3. B.
5. B.
7. A.
9. A.

SECTION 7, ANSWERS AND SOLUTIONS

PART A
ELECTRICAL CODE

1. **A.** 80 A divided by 2 = 40 A $\times$ 125% = 50 A.
   B. No. 6 R90 X-Link CU.
3. 33 A divided by 2 = 16.5 A $\times$ 115% = 18.98 A.

PART B
TROUBLESHOOTING

1. If the conductor on the left side of the TR coil is broken and the start button is pressed, the S coil would become energized, closing the S contact in the control circuit, which will maintain the circuit. The three S contacts will close, which will energize the first part of the motor winding. The motor will remain on the first stage.

3. If the N/O S contact is defective and will not close and the start button is pressed momentarily, the S and TR coils will be energized momentarily. The circuit would shut down when the start button is released.

5. If the N/C R contact were defective and will not close and the start button is pressed, only the S coil would be energized. The N/O S contact will close to maintain the circuit. The TR coil will not be energized. Nothing more will happen in the control circuit. The motor will continue to operate on the first stage.

7. If the S coil has an open circuit and the start button is pressed, nothing will happen.

PART C
MULTIPLE-CHOICE QUESTIONS

1. C.
3. D.
5. A.
7. C.

SECTION 8, ANSWERS AND SOLUTIONS

PART A
ELECTRICAL CODE

1. **A.** 60 A × 125% = 75 A.
   B. No. 4 R90 X-Link CU.
3. 41 A × 125% = 51.25 A.

PART B
TROUBLESHOOTING

1. The motor will operate only while the start button is held closed. Releasing the start button will cause the circuit to shut down.

3. In the normal operation of the control circuit, when the N/C R contact opens, the TR coil will deenergize, opening the TR–TC contact. If because of a fault, the N/C R contact does not open, the TR coil will remain energized. This will have no effect on the operation of the circuit.

5. If because of a fault, the three R power contacts closed before the S power contacts, the motor would start on full line voltage, rather than reduced voltage. The starting

current would be greater than it would if started on reduced voltage.

7. If because of a fault, the TR–TC contact will not close, pressing the start button will energize the S and TR coils. The N/O S (control) contact will close, maintaining the circuit. The S power contacts will close at the same time, connecting the resistance units in series with the motor windings. The resistors will remain in the circuit and eventually overheat.

PART C
MULTIPLE-CHOICE QUESTIONS

1. B.
3. B.
5. C.
7. D.
9. D.

SECTION 9, ANSWERS AND SOLUTIONS

PART A
ELECTRICAL CODE

1. A. 68 A divided by $\sqrt{3} \times 125\% = 49.07$ A

 B. No. 8 R90 X-Link CU.

3. 68 A $\times$ 175% = 119 A. Therefore select a 125-A fuse.

PART B
TROUBLESHOOTING

1. The TR coil will never become energized, which in turn will never deenergize the S coil. As a result, the motor will be left in the starting (wye) configuration.

3. The M, TR, and S coils will become energized, and following the time delay, the S coil will not be deenergized, leaving the M and S power contacts closed. The motor will be left running in the wye configuration.

5. The TR coil will become energized and one TR contact will maintain the TR coil circuit. The other coils will not become energized. The motor will not start.

PART C
MULTIPLE-CHOICE QUESTIONS

1. A.
3. C.
5. C.
7. B.
9. A.
11. C.
13. B.
15. A.

SECTION 10, ANSWERS AND SOLUTIONS

PART A
ELECTRICAL CODE

1. **A.**   75 A × 125% = 93.75 A.

    B.   No. 3 R90 X-Link CU.
3.   70 A × 150% = 105 A.

Therefore, select a 110-A time-delay fuse.

PART B
TROUBLESHOOTING

1.   The R coil will remain deenergized, causing the resistors to remain in the rotor circuit and the motor to run at a reduced speed.

3.   The slip-rings are bypassed, causing a high starting current in the primary winding.

PART C
MULTIPLE-CHOICE QUESTIONS

1. D.
3. A.
5. D.
7. A.
9. D.

SECTION 11,
ANSWERS
AND SOLUTIONS

PART A
ELECTRICAL CODE

1. 68 A × 125% = 85 A.
3. No. 3 R90 X-Link CU.
5. 1/2 in. (13 mm) conduit.

PART B
TROUBLESHOOTING

1. The FLR coil would be deenergized, opening the FLR contact, deenergizing the M coil, stopping the motor from running away.
3. The motor would stop.
5. A short-circuit would occur when the M power contacts closed.
7. The motor would start, but will shut down when the start button is released.

PART C
MULTIPLE-CHOICE QUESTIONS

1. B.
3. C.
5. D.
7. A.
9. D.
11. C.

GLOSSARY OF TERMS

ABOVE-BASE SPEED   This speed is achieved by inserting resistance in the shunt field to operate above the rated base speed of a DC motor.

ACROSS-THE-LINE STARTING   A motor is started on full line voltage with no provision for restricting the starting current.

AMPERE(A)   Unit of current flow.

ANTI-PLUGGING   A circuit or installation that prevents the motor from being reversed without first allowing the motor to stop, or that will not allow the motor windings to be utilized for braking.

AUTOMATIC CONTROLLER   A motor starter that may be used to control a motor automatically.

AUTOTRANSFORMER   A transformer, consisting of a single winding, wound on a laminated iron core. Percentage taps are extended to obtain various starting voltages.

AUTOTRANSFORMER STARTING   A starting method used to start large squirrel-cage motors that restricts starting current and controls starting torque.

BASE SPEED   The speed of a DC motor when the motor is properly connected to rated voltage.

BELOW-BASE SPEED   Achieved by inserting resistance in the armature circuit to operate below the rated speed of a DC motor.

BIMETAL OVERLOAD RELAY   Comprised of an overload heater element adjacent to a bimetal strip. The bimetal strip is composed of two dissimilar metals, fused together. When the strip is heated, it will bend in a predetermined direction, tripping the motor starter, stopping the motor.

BLOWOUT COIL   A low resistance coil on some DC motor starters, connected in series with the main power contacts to extinguish the arc.

BRANCH CIRCUIT   That portion of an electrical circuit beyond the final overcurrent device.

CLOSED TRANSITION   A motor is not disconnected from the line during the start-up period when switching from one step to another.

CODE   The (electrical) code is the safety standard by which the electrical industry is governed.

COIL SEQUENCE   A method of listing the order in which magnetic coils are energized or deenergized.

COMPENSATOR STARTER   Another name given to an autotransformer motor starter.

COMPOUND MOTOR   A DC motor consisting of a shunt winding and a series winding in addition to the armature.

CONTACT   Equivalent to a switch. A contact in a motor starter or relay is controlled by a magnetic coil.

CONTACTOR   An electrically operated switch (coil) that is generally rated in amperes.

CONTROL CIRCUIT   That portion of the electrical installation that contains the coils, solenoids, and overload contacts, and the control device that initiates the operation of the equipment.

CONTROLLER   Another name for motor starter.

CONTROL RELAY   A name given to a magnetic contactor in the smaller sizes.

CONTROL TRANSFORMER   The transformer used to lower the line voltage to a lower value for the control circuit.

CEMF STARTER   (counter electromotive force) A method of starting a DC motor, using the CEMF, to determine the starting time.

CURRENT (A)   Unit of current flow.

DASHPOT ACCELERATION A method of starting a DC motor, using an oil dashpot to determine the starting time, which will restrict the starting current.

DISCONNECTING MEANS Any device that will permit the motor to be safely disconnected from the supply voltage.

DRUM SWITCH A manually operated three-position three-pole rotary switch, which carries a hp rating, and is used for manually reversing AC or DC motors.

DUAL VOLTAGE MOTOR A motor designed to operate on two different voltages. The motor connections must be altered to suit the supply voltage.

DUAL VOLTAGE RELAY A relay consisting of a coil and contacts. A barrier separates the two components.

DYNAMIC BRAKING A method of braking that uses the DC motor as a generator during the braking period.

EEMAC Represents the Canadian counterpart of NEMA used in the United States. Electrical and Electronic Manufacturers Association of Canada.

ELECTRICAL CODE The safety standard by which the electrical industry is governed.

ELECTRICAL INTERLOCK A contact connected in a control circuit that will insure that a particular sequence of operation is followed.

ELECTROMAGNET Consists of a coil of wire wound on an iron core.

ELEMENTARY DIAGRAMS (See Schematic Diagram.)

FACEPLATE RESISTOR UNIT A bank of resistors with a set of contacts manually controlled to adjust the resistance values.

FOUR-POINT STARTER A motor starter consisting of four connection points. Used to start a DC motor. The starter restricts starting current and controls the starting torque.

FULL VOLTAGE STARTING This term means that the motor is started on full line voltage with no provision for restricting the starting current.

HOLDING COIL (See Magnetic Coil.)

HOLDING CONTACT (See Maintaining Contact.)

INCHING (See Jogging.)

JOGGING An operation in which the motor runs when the pushbutton is pressed and will stop when the pushbutton is released.

LADDER DIAGRAMS   (See Schematic Diagram.)

LINE DIAGRAMS   (See Schematic Diagram.)

LOCK-OUT RELAY   Used in a current-limit starter for starting a DC motor and is dependent on armature current to operate.

LONG SHUNT   A compound motor connection in which the shunt field is connected across the series combination of the series field and the armature of a DC motor.

LOW VOLTAGE PROTECTION   When the voltage for a magnetic coil is low, the contacts will open, causing the motor to stop. When the voltage is restored, the motor will not re-start automatically.

LOW VOLTAGE RELEASE   When the voltage for a magnetic coil is low, the contacts will open, causing the motor to stop. When the voltage is restored, the motor will restart automatically.

MAGNETIC COIL   The term given to the coil of wire on an electromagnet.

MAGNETIC CONTACTOR   (See Contactor.)

MAGNETIC MOTOR STARTER   A motor starter that may be controlled from a remote location by energizing an electromagnetic coil.

MAINTAINING CONTACT   A normally open contact located generally in the magnetic motor starter. Its function is to bypass the momentary start pushbutton.

MANUAL MOTOR STARTER   A horsepower rated switch for starting a motor manually. The motor starter will have provision for overload protection.

MECHANICAL INTERLOCK   A mechanism that is factory-installed, which will prevent two sets of contacts from closing simultaneously.

MOTOR CONTROL CENTER   A cabinet containing more than one motor starter.

MOTOR SERVICE FACTOR   Means that the motor may be allowed to develop more than its rated hp without causing undue deterioration of the insulation. The service factor is a margin of safety.

NEMA   Represents the American counterpart of EEMAC used in Canada. National Electrical Manufacturers Association.

NO FIELD RELEASE   A system designed to protect a DC motor from racing out of control above base speed.

NO VOLTAGE PROTECTION   A motor will stop when there is a supply voltage failure and will not restart automatically when the supply voltage is restored.

NO VOLTAGE RELEASE   A motor will stop when there is a supply voltage failure and will restart automatically when the supply voltage is restored.

OPEN TRANSITION   A motor is disconnected from the line during the start-up period when switching from one step to another.

OVERCURRENT PROTECTION   Is installed in the circuit to protect the conductors and may be in the form of fuses or circuit breakers.

OVERLOAD CONTACT   A contact connected in the coil circuit of a motor starter that is held closed by the overload relay mechanism. If an overload should persist, the contact will open, deenergizing the coil, stopping the motor.

OVERLOAD PROTECTION   Is installed in the power circuit to protect the motor and may be in the form of relays, heaters, or elements.

PART-WINDING MOTOR STARTING   A reduced current method of starting for squirrel-cage motors that have two separate stator windings connected in parallel.

PERCENTAGE TAPS   The taps extended from an autotransformer.

PILOT DEVICE   A control device that is capable of initiating a signal to complete a control circuit (for example, thermostat).

PLUGGING   A method of stopping a polyphase motor quickly by momentarily connecting the motor for the reverse rotation when the motor is running.

POWER CIRCUIT   That portion of the electrical circuit that serves to supply the stator winding of the motor.

PRIMARY CIRCUIT   The stator winding is referred to as the primary of a motor.

PRIMARY RESISTANCE STARTING   A reduced voltage method of starting squirrel-cage motors. The starting method uses resistance units in the primary circuit to restrict starting current.

REDUCED CURRENT STARTING   Any starting method that restricts starting current is referred to as reduced current starting.

REDUCED VOLTAGE STARTING   Any starting method that reduces starting voltage to restrict starting current is referred to as reduced voltage starting.

RELAY   (See Control Relay.)

REMOTE STATION   The station or control is remote from the magnetic motor starter.

REVERSING MAGNETIC MOTOR STARTING   A motor starter capable of reversing a motor and may be controlled from a remote location.

RING CIRCUIT   A circuit in which the motors start in sequence and the maintaining contact of the last motor starter seals the circuit.

SEAL-IN CONTACT   (See Maintaining Contact.)

SCHEMATIC DIAGRAM   The control components are rearranged to simplify the tracing of the circuit.

SECONDARY RESISTANCE MOTOR STARTING   A reduced current method of starting a wound rotor motor. The starting method uses resistance units in the rotor circuit.

SEQUENCE CONTROL   When a multiple motor installation must operate in a predetermined sequence.

SERIES MOTOR   A DC motor having a series field and armature.

SHADING RING (Coil)   A single turn of conducting material mounted in the face of the magnetic assembly or armature electromagnet used on alternating current.

SHORT SHUNT   A compound motor connection in which the shunt field is connected across the armature.

SHUNT MOTOR   A DC motor having a shunt field and armature.

SLIP-RING MOTOR   (See Wound Rotor Motor.)

STAR-DELTA MOTOR STARTING   A reduced current method of starting a squirrel-cage motor that has its windings connected star for starting and delta for running.

THREE-POINT STARTER   A motor starter consisting of three connection points. Used to start a DC motor. The starter restricts starting current and controls starting torque.

TRANSFORMER (See Control Transformer, see Auto-transformer.)

TRANSITION (See Closed Transition, Open Transition.)

TWO-COIL LOCK-OUT RELAY (See Lock-Out Relay.)

UNDERVOLTAGE PROTECTION (See Low Voltage Protection.)

UNDERVOLTAGE RELEASE (See Low Voltage Release.)

WIRING DIAGRAM Includes all of the devices in the system and shows their physical relationship to each other.

WOUND ROTOR MOTOR Consists of a squirrel-cage stator winding and a wound rotor, terminating on three slip-rings, which will permit connections to external resistance units.

WYE-DELTA MOTOR STARTING (See Star-Delta Motor Starting.)

ZERO-SPEED SWITCH Monitors rotation of rotating equipment. May be used for plugging and anti-plugging installations.

INDEX

CONTENTS

LAB SECTION A
AC MOTOR CONTROL LABS

ZERO-SPEED SWITCH

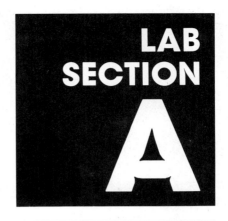

LAB SECTION A

AC MOTOR CONTROL LABS

INTRODUCTION TO AC LABS

The Motor Control Lab section has been prepared to give the student an opportunity to apply the information covered in the text in a "hands-on" everyday situation. It has been designed so that a student may progress at his or her own speed.

The teacher may elect to assign labs that coincide with the classroom lectures. If a student is able to relate the classroom lectures to the assigned labs, much learning should result.

MOTOR CONTROL WORKSTATION

The motor control workstation shown in Figure A-1 is the suggested layout. A layout has been provided so that the physical layout illustrated for each lab will relate to an actual installation.

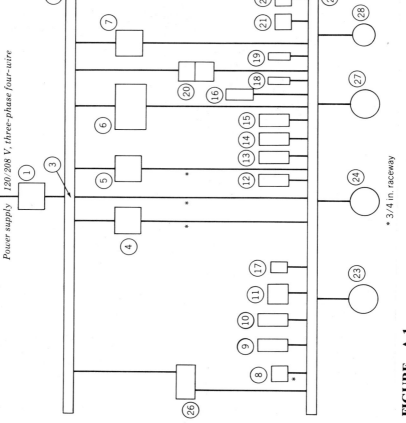

FIGURE A-1
Motor Control Workstation.

1. 15 A Three-Pole circuit breaker.
2. 4 in. × 4 in. wiring trough.
3. Power supply terminal strip.
4. Magnetic across-the-line motor starter.
5. Magnetic across-the-line motor starter.
6. Reversing magnetic motor starter.
7. Manual three-phase across-the-line motor starter.
8. Control relay.
9. Single-pole thermostat.
10. Single-pole manual motor starter.
11. Pneumatic timing relay.
12. Start–stop remote pushbutton station.
13. Start–stop remote pushbutton station.
14. Start–jog–stop remote pushbutton station.
15. Start–stop remote pushbutton station with a single-pole selector switch.
16. Forward–reverse–stop remote pushbutton station.
17. Single-pole toggle switch.
18. Limit switch no. 1.
19. Limit switch no. 2.
20. Dual voltage relay.
21. Pilot light.
22. Hand-off-auto selector switch.
23. Three-phase 1/4 HP motor.
24. Split-phase 1/4 HP motor.
25. 4 in. × 4 in. wiring trough.
26. Control transformer primary-208 V, secondary-120 V.
27. Three-phase 1/4 HP motor.
28. Zero-speed switch.

LAB INSTRUCTIONS

Each lab has been divided into several parts:

Physical layout.
Wiring diagram.
Schematic diagram.
Instructions.
Conclusions.

Physical Layout

The physical layout relates to the motor control workstation, shown in Figure A-1. This layout should assist the student to relate to a motor control installation.

Wiring Diagram

The wiring diagram (as mentioned in the text) relates to the actual position of each component of the control system.

Schematic Diagram

The schematic diagram will match the wiring diagram, but will be easier to read.

Instructions

The instructions listed for each lab will assist the student in obtaining as much information as possible from each lab.

Conclusions

The conclusions will include a summing up of the lab and will have comments on the electrical code if applicable.

Evaluations of the Labs

The teacher may evaluate each lab at the workstation or request lab reports to be submitted for evaluation.

Answers, conclusions, and completed drawings for all labs will be located in the Instructor's Manual.

Lab Layout

Physical layouts have been provided for each lab, as well as wiring diagram layouts for the first eleven labs and schematic diagram layouts for the first four labs. Instructions have also been supplied for all labs.

General instructions apply only to labs that require the student to install the wiring and complete the electrical connections.

GENERAL INSTRUCTIONS FOR LAB NOS. 1 TO 20

1. Indicate the number of conductors required and the colors chosen on the physical layout provided for each lab.
2. Complete the wiring diagram, showing the colors to match the physical layout.
3. Complete the schematic diagram, also showing the colors to match the physical layout.
4. Install the conductors. Complete all connections. Operate the equipment.
5. Perform the requirements listed for each lab.
6. Record any relevant data.
7. Identify all parts of the circuits.
8. Answer all questions.
9. Wiring to be installed in a neat and professional manner.
10. Do not abuse the equipment.
11. Have the installation and conclusions evaluated by the teacher.

LAB NO. 1
SINGLE-POLE MANUAL
MOTOR STARTER TO OPERATE
A 120-V SPLIT-PHASE MOTOR

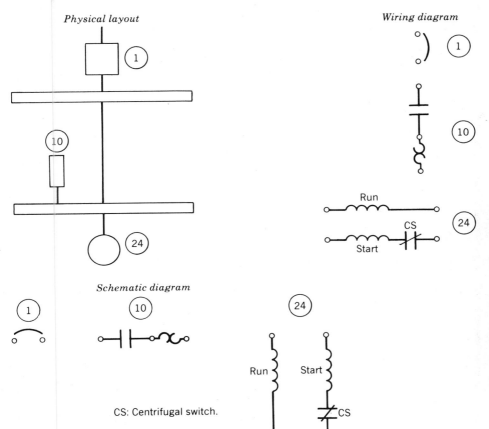

Physical layout

Wiring diagram

Schematic diagram

CS: Centrifugal switch.

INSTRUCTIONS

1. Follow the general instructions.
2. Select the correct overload heater for the installation.
3. Reverse the direction of rotation of the split-phase motor.
4. Is the installation a NVR or a NVP circuit?

CONCLUSIONS

LAB COMPLETED _____   **TEACHER** _____

LAB NO. 1

LAB NO. 2
THREE-POLE
MANUAL MOTOR STARTER TO
OPERATE A THREE-PHASE MOTOR

Physical layout

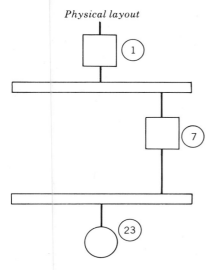

Wiring diagram

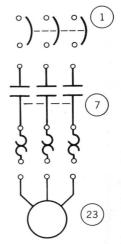

Schematic diagram

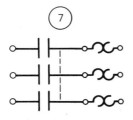

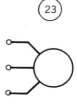

INSTRUCTIONS

1. Follow the general instructions.
2. Select the correct overload heater for the installation.
3. Reverse the direction of rotation of the three-phase motor.
4. Is the installation a NVR or a NVP circuit?

CONCLUSIONS

LAB COMPLETED _____ **TEACHER** _____

LAB NO. 3
MAGNETIC
ACROSS-THE-LINE
MOTOR STARTER TO
OPERATE A THREE-PHASE
MOTOR; TWO-WIRE CONTROL

Physical layout

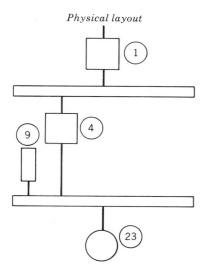

Wiring diagram

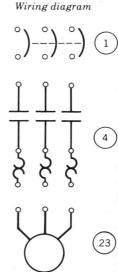

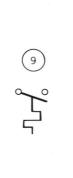

Schematic diagram

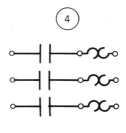

LAB NO. 3

INSTRUCTIONS

1. Follow the general instructions.
2. Is the installation a NVR or a NVP circuit?
3. Prove the answer to no. 2 based on the circuit operation and the electrical code.

CONCLUSIONS

LAB COMPLETED _____ **TEACHER** _____

© 1987 by John Wiley & Sons, Inc.

LAB NO. 3

LAB NO. 4
MAGNETIC
ACROSS-THE-LINE
MOTOR STARTER TO
OPERATE A THREE-PHASE
MOTOR; THREE-WIRE CONTROL

Physical layout

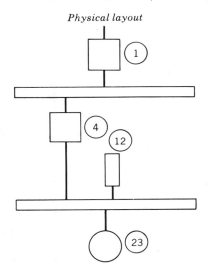

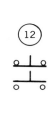

Wiring diagram

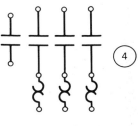
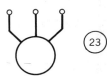

Schematic diagram

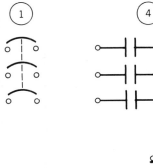
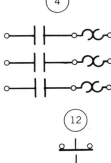

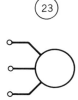

LAB NO. 4

INSTRUCTIONS

1. Follow the general instructions.
2. Is the installation a NVR or a NVP circuit?
3. Prove the answer to no. 2 based on the circuit operation and the electrical code.

CONCLUSIONS

LAB NO. 4

LAB COMPLETED _____ **TEACHER** _____

LAB NO. 5
CONTROL CIRCUIT,
THREE-WIRE CONTROL; TWO-POINT
CONTROL (POWER CIRCUIT LAB NO. 3)

Physical layout

Wiring diagram

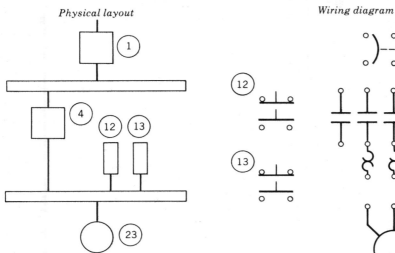

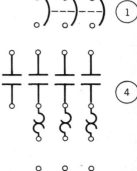

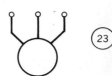

Schematic diagram

© 1987 by John Wiley & Sons, Inc.

LAB NO. 5

LAB NO. 5

INSTRUCTIONS

1. Follow the general instructions.
2. Is the installation a NVR or a NVP circuit?
3. State the rule to be remembered when connecting multiple start–stop pushbuttons.

CONCLUSIONS

LAB COMPLETED _____ **TEACHER** _____

© 1987 by John Wiley & Sons, Inc.

LAB NO. 6
JOG CONTROL CIRCUIT.
START–STOP PUSHBUTTON
STATION WITH A SELECTOR SWITCH
TO ALLOW THE START PUSHBUTTON
TO ALSO FUNCTION AS A JOG
BUTTON (POWER CIRCUIT LAB NO. 3)

Physical layout

Wiring diagram

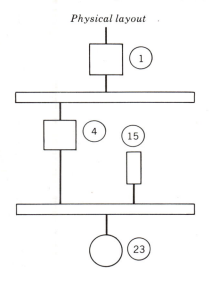

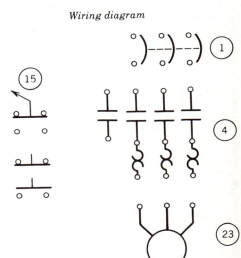

Schematic diagram

© 1987 by John Wiley & Sons, Inc.

INSTRUCTIONS

1. Follow the general instructions.
2. Is the installation a NVR or a NVP circuit?
3. Observe circuit operation and record.
4. Record the disadvantage of this control circuit.

CONCLUSIONS

LAB COMPLETED _____ **TEACHER** _____

© 1987 by John Wiley & Sons, Inc.

LAB NO. 7
JOG CONTROL CIRCUIT.
START–JOG–STOP PUSHBUTTON
STATION (POWER CIRCUIT LAB NO. 3)

Physical layout

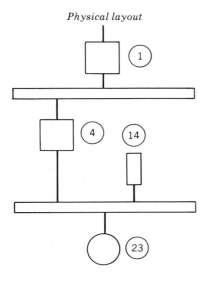

Wiring layout

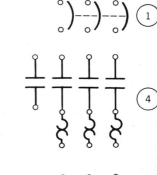

Schematic diagram

LAB NO. 7

INSTRUCTIONS

1. Follow the general instructions.
2. Observe the circuit operation and record.
3. Explain why this circuit could be hazardous.

CONCLUSIONS

LAB COMPLETED _____ **TEACHER** _____

© 1987 by John Wiley & Sons, Inc.

LAB NO. 8
JOG CONTROL CIRCUIT.
START–JOG–STOP PUSHBUTTON
STATION WITH A JOG RELAY ADDED TO
PROVIDE SAFE OPERATION OF THE
EQUIPMENT (POWER CIRCUIT LAB NO. 3)

Physical layout

Wiring diagram

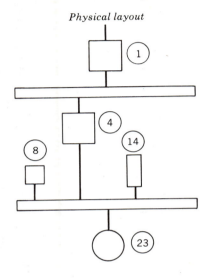

Schematic diagram

LAB NO. 8

INSTRUCTIONS

1. Follow the general instructions.
2. Observe the circuit operation and record.
3. Record the coil sequence.

CONCLUSIONS

LAB COMPLETED _____ **TEACHER** _____

© 1987 by John Wiley & Sons, Inc.

LAB NO. 9
CONTROL TRANSFORMER INSTALLED TO PROVIDE 120-V SUPPLY TO OPERATE A START–STOP CONTROL CIRCUIT (POWER CIRCUIT LAB NO. 3)

Physical layout

Wiring diagram

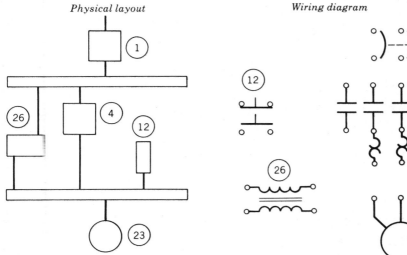

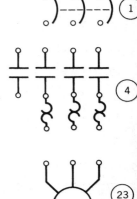

Schematic diagram

© 1987 by John Wiley & Sons, Inc.

INSTRUCTIONS

1. Follow the general instructions.
2. Ground the secondary winding of the transformer correctly.
3. Start the motor.
4. Place a ground fault on the control circuit. Observe and record results.
5. Ground the secondary winding of the transformer incorrectly.
6. Start the motor.
7. Place a ground fault on the control circuit. Observe and record results.

CONCLUSIONS

LAB COMPLETED _____ **TEACHER** _____

LAB NO. 10
HAND-OFF-AUTO CONTROL
CIRCUIT WITH A SINGLE-POLE
THERMOSTAT (POWER CIRCUIT LAB NO. 3)

Physical layout

Wiring diagram

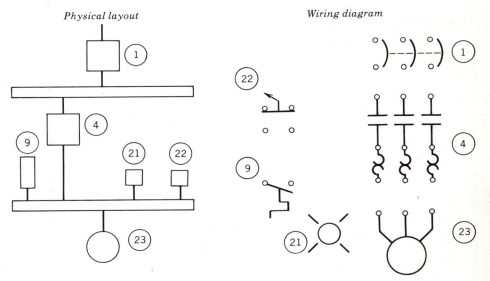

Schematic diagram

LAB NO. 10

INSTRUCTIONS

1. Follow the general instructions.
2. Observe circuit operation.
3. Is the installation a NVR or a NVP circuit?
4. Install a pilot light to indicate when the magnetic coil is energized.

CONCLUSIONS

LAB COMPLETED _____ **TEACHER** _____

© 1987 by John Wiley & Sons, Inc.

LAB NO. 11
POWER AND CONTROL
CIRCUIT FOR A REVERSING
MAGNETIC MOTOR STARTER,
SIZE 00 (Mechanical Interlock Only)

Physical layout

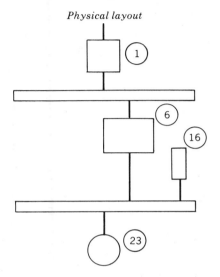

Schematic diagram

Wiring diagram

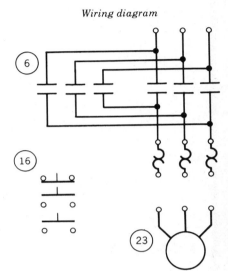

LAB NO. 11

INSTRUCTIONS

1. Follow the general instructions.
2. Observe circuit operation.
3. Locate and examine the mechanical interlock in the reversing magnetic motor starter.
4. Explain the purpose for and the operation of the mechanical interlock.

CONCLUSIONS

LAB COMPLETED _____   **TEACHER** _____

LAB NO. 11

LAB NO. 12
CONTROL CIRCUIT FOR A
REVERSING MAGNETIC MOTOR
STARTER (ELECTRICAL
INTERLOCK IN THE MOTOR
STARTER, POWER CIRCUIT LAB NO. 11)

Physical layout

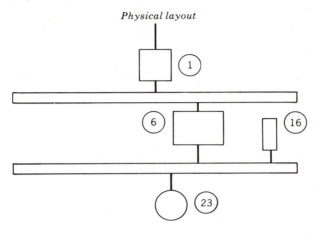

Schematic diagram

© 1987 by John Wiley & Sons, Inc.

INSTRUCTIONS

1. Follow the applicable general instructions.
2. Observe circuit operation.
3. Explain the purpose for and the operation of the electrical interlocks in the reversing magnetic motor starter.

CONCLUSIONS

LAB COMPLETED _____ **TEACHER** _____

© 1987 by John Wiley & Sons, Inc.

LAB NO. 12

LAB NO. 13
CONTROL CIRCUIT FOR A REVERSING MAGNETIC MOTOR STARTER (ELECTRICAL INTERLOCKS IN THE REMOTE PUSHBUTTONS AND THE MOTOR STARTER, POWER CIRCUIT LAB NO. 11)

Physical layout

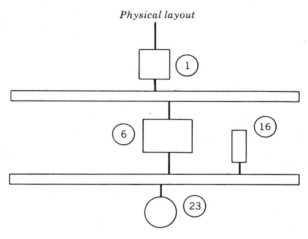

Schematic diagram

© 1987 by John Wiley & Sons, Inc.

LAB NO. 13

INSTRUCTIONS

1. Follow the applicable general instructions.
2. Observe the circuit operation.
3. Write and submit a report on the operation of the complete control and power circuits.

CONCLUSIONS

LAB COMPLETED _____ **TEACHER** _____

© 1987 by John Wiley & Sons, Inc.

LAB NO. 14
INTERLOCKING
TWO THREE-PHASE
MOTORS AS PER INSTRUCTIONS

Physical layout

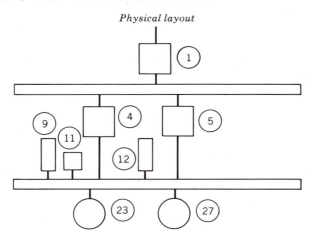

Schematic diagram

INSTRUCTIONS

1. Follow the applicable general instructions.
2. The motors must operate as follows:
   A. Pressing the start button will start the first motor.
   B. After a ten-second time delay, the second motor will start (if the thermostat contact is in the closed position).
3. Use this circuit to practice troubleshooting.

CONCLUSIONS

LAB COMPLETED _____ **TEACHER** _____

© 1987 by John Wiley & Sons, Inc.

LAB NO. 15
SEQUENCE CONTROL—RING CIRCUIT

Physical layout

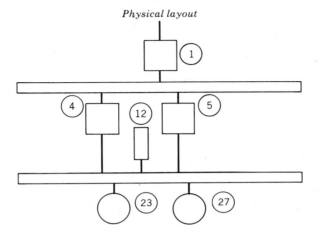

Schematic diagram

INSTRUCTIONS

1. Follow the applicable general instructions.
2. The motors must operate as follows:
    A. Pressing the start button will start the first motor.
    B. The second motor will start.
    C. If an overload contact should open, the motors will stop in sequence.
    D. If we press the stop button, the motors will also stop in sequence.
3. Use this circuit to practice troubleshooting.

CONCLUSIONS

LAB COMPLETED _____ **TEACHER** _____

LAB NO. 16
SEMIAUTOMATIC CONTROL
CIRCUIT, LIMIT-SWITCHES,
AND A PNEUMATIC TIMER

Physical layout

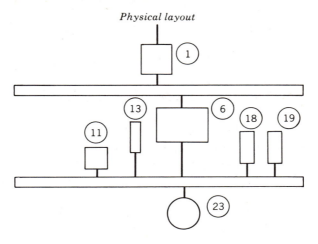

Schematic diagram

INSTRUCTIONS

1. Follow the applicable general instructions.
2. The circuit must operate as follows:
   A. Pressing the start button allows the motor to run in the forward rotation.
   B. Tripping the limit-switch no. 1 stops the forward rotation.
   C. Following a ten-second delay, the motor starts in the reverse rotation.
   D. Tripping the limit-switch no. 2 stops the motor.
   E. *Note:* Do not release the no. 1 limit-switch until the motor reverses rotation.

CONCLUSIONS

LAB COMPLETED _____ **TEACHER** _____

ZERO-SPEED SWITCH

Lab nos. 17 through 20 deal with plugging and anti-plugging. A zero-speed switch is required to perform both functions. Item no. 28 on the work station is a zero-speed switch. Such a switch is very costly, and if budget restraints will not permit such an expenditure, then possibly one or two of the worksta-tions could be so equipped.

LAB NO. 17
PLUGGING A MOTOR
FROM ONE DIRECTION ONLY

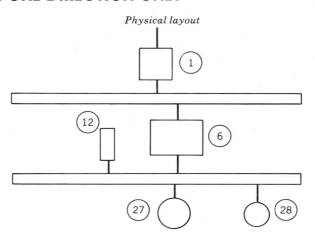

Physical layout

Schematic diagram

See Fig. 4-21

INSTRUCTIONS

1. Follow applicable general instructions.
2. Operate the equipment.
3. Using a digital "clip-on" ammeter, record
   A. Starting current.
   B. Running current.
   C. Plugging current.
4. Relate the respective current readings to a larger motor.
5. Keep in mind that this circuit could be hazardous.

CONCLUSIONS

LAB COMPLETED _____ **TEACHER** _____

LAB NO. 18
PLUGGING A MOTOR
FROM ONE DIRECTION ONLY

Physical layout

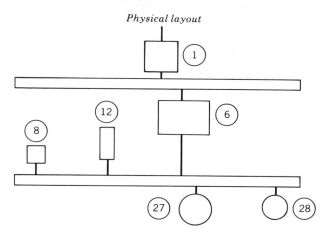

Schematic diagram
See Fig. 4-23

LAB NO. 18

INSTRUCTIONS

1. Follow the applicable general instructions.
2. Operate the equipment.
3. Study the circuit operation to understand fully why the circuit is safer than that in Lab no. 17.

CONCLUSIONS

LAB COMPLETED _____ **TEACHER** _____

LAB NO. 18

LAB NO. 19
PLUGGING A MOTOR
FROM BOTH DIRECTIONS

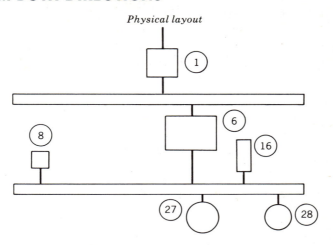

Physical layout

Schematic diagram
See Fig. 4-27

LAB NO. 19

LAB NO. 19

INSTRUCTIONS

1. Follow the applicable general instructions.
2. Operate the equipment.
3. *Note:* The zero-speed switch for this control circuit must be equipped with a built-in lock-out coil.

CONCLUSIONS

LAB COMPLETED _____ **TEACHER** _____

LAB NO. 20
ANTI-PLUGGING

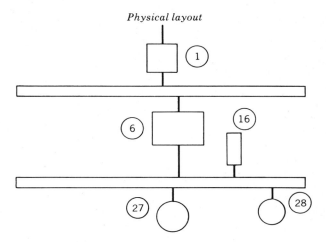

Physical layout

Schematic diagram
See Fig. 4-29

INSTRUCTIONS

1. Follow the applicable general instructions.
2. Operate the equipment.
3. Submit a report explaining how the circuit operates and why, on a large motor, anti-plugging should be used.

CONCLUSIONS

LAB COMPLETED _____ **TEACHER** _____

© 1987 by John Wiley & Sons, Inc.

LAB NO. 20

INTRODUCTION TO LAB NOS. 21 TO 26

Labs 21 through 26 have been prepared to help the student understand the operation of the various methods used to reduce the starting current of large motors. The text explains the purpose for and the operation of each starting method.

It would be an impossible task to analyze and run tests on each motor control circuit in use at the present time, but the knowledge gained from these labs can be easily adapted to other circuits. The Allen-Bradley and Square D motor control companies have kindly consented to allow their circuits to be used in this text. The teacher could easily adapt the lab instructions to suit his or her particular installed equipment.

Workstation Requirements

Reduced voltage or reduced current starting is required to start large motors. Due to budget and space restraints, most electrical labs are not equipped with large test motors.

The next best thing is to obtain from your supplier the smallest motor starter available for the particular starting method to be installed in each workstation. It is also important to install motor starters matching in size to the respective motor.

Conclusions and Evaluation

The text and the lab portion have been designed to assist the student to research and prepare for labs outside lab periods. Never treat any piece of motor control equipment as a black box, full of the unknown. Everything in a piece of control equipment is important and should be understood.

The student must be encouraged to keep good notes and prepare neat drawings. This will enable the student to demonstrate or discuss the topics in a professional manner.

Keep in mind that the application of the electrical code to the motor installation is important.

Troubleshooting

Labs 21 to 26 lend themselves very well to troubleshooting procedures. Under normal conditions in industry, the fault has occurred, and the control and power circuits operate in an improper manner. The reverse approach will be taken when troubleshooting equipment in the lab.

Faults will be ascribed to the circuits and the problem will be to determine how the equipment will function. A close study of the schematic diagrams should reveal the answers. Some suggested faults include

1. Disconnect a conductor from any point in the circuit.
2. Bypass a contact.
3. The students should consult with one another so that each will know what function is being performed.
4. Discuss the results of each step so that as much information is obtained from each lab as possible.
5. If a student is uncertain about the lab results, consult the text for assistance.
6. Good lighting is essential around all equipment.
7. Think *safety* at all times.

Tools Required

The tools required to perform the labs will be

1. Voltmeter (two).
2. Ohmmeter.
3. Clip-on ammeter.
4. Flat, medium screwdriver.
5. Tachometer.

LAB NO. 21
PRIMARY
RESISTANCE STARTING—MANUAL

Physical layout

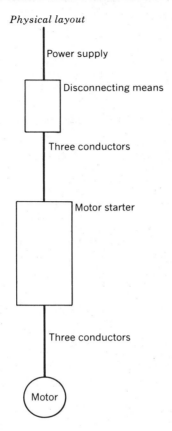

Power supply

Disconnecting means

Three conductors

Motor starter

Three conductors

Motor

Power circuit-wiring diagram
See Fig. 8-15
Control circuit-schematic diagram
See Fig. 8-15

INSTRUCTIONS

1. Lock the disconnecting means in the off position.

2. Record the motor starter ratings and the motor data.

3. Study the construction and the general layout of the motor starter.

4. Using the schematic diagram, locate every terminal, contact, resistor, overload device, and coil in the motor starter.

5. Using the ohmmeter, measure the starting resistance. Close the starting contact and slowly raise the handle to the top position. Note the effect on the resistance.

6. Is the pickup contact N/O of N/C? Be certain of your answer.

7. The starting current does not pass through the overload heaters. Is this acceptable by the electrical code? Prove your answer.

8. Unlock the disconnecting means and start the motor following the instructions given on the motor starter cover.

9. Monitor the starting current and starting voltage for varying starting times.

10. Is the installation NVR or NVP?

11. Is the transition open or closed?
12. Is the starting method reduced voltage or reduced current?

CONCLUSIONS

LAB COMPLETED _____ TEACHER _____

© 1987 by John Wiley & Sons, Inc.

LAB NO. 22
TWO-PART-WINDING STARTING

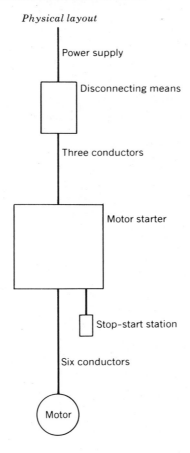

Physical layout

Power supply

Disconnecting means

Three conductors

Motor starter

Stop-start station

Six conductors

Motor

Power circuit—wiring diagram

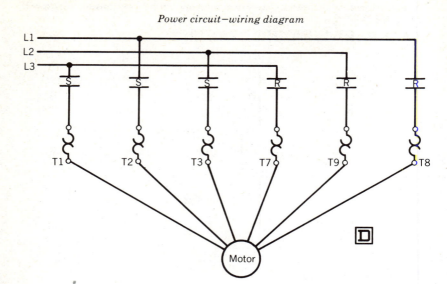

Control circuit—schematic diagram

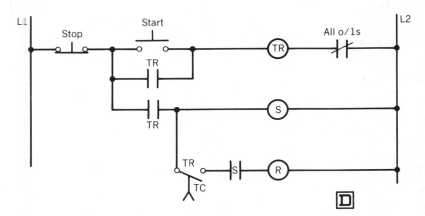

INSTRUCTIONS

1. Lock the disconnecting means in the off position.
2. Record the motor starter ratings and the motor data.
3. Study the general layout of the interior of the motor starter.
4. Using the schematic diagram, locate every terminal, contact, overload device, and coil in the motor starter.
5. Note two sets of overload heaters. Select the correct overload heaters for this installation.
6. Adjust the time delay for three seconds.
7. Unlock the disconnecting means.
8. Start the motor.
9. Measure the running current at:
   A. L1_____A.
   B. T1_____A.
   C. T7_____A.
10. Extend the time delay to approximately 60 seconds.
11. Start the motor.
12. Measure the voltage across the:
    A. T1 and T2 terminals_____.
    B. T7 and T8 terminals_____(before the preset time delay completes the run circuit).
    Explain the reason for the readings.
13. Is the time-delay contact released by a TR coil, or is it released (triggered) by another contactor?
14. Measure:
    A. Starting current at L1_____.
    B. Changeover current at L1_____.

15. Is part-winding starting open or closed transition?

16. Is this circuit NVR or NVP?

17. List the coil sequence.

18. When the complete circuit is fully understood, attempt to troubleshoot the circuit.

For troubleshooting information, see introduction to Lab nos. 21 to 26, ''Troubleshooting.''

CONCLUSIONS

LAB COMPLETED _____ **TEACHER** _____

LAB NO. 22

LAB NO. 23
WYE-DELTA STARTING

Physical layout

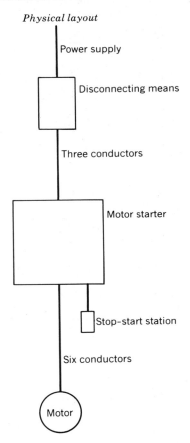

Power supply

Disconnecting means

Three conductors

Motor starter

Stop-start station

Six conductors

Motor

Power circuit–schematic diagram

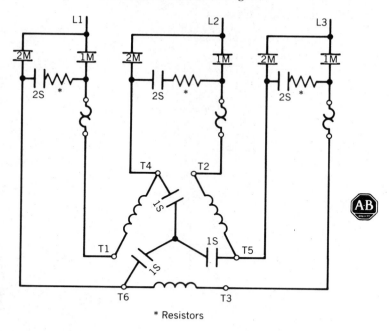

* Resistors

Control circuit—schematic diagram

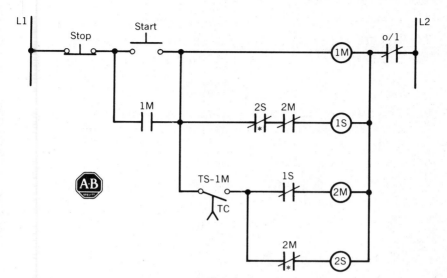

* Late-break contact

INSTRUCTIONS

1. Lock the disconnecting means in the off position.
2. Record the motor starter ratings and the motor data.
3. Study the general layout of the interior of the motor starter.
4. Using the schematic diagram, locate every terminal, contact, overload device, and coil in the motor starter.
5. Note where the overload heaters are installed in the power circuit. Select the correct overload heater for this installation.
6. Adjust the time delay for eight seconds.
7. Unlock the disconnecting means.
8. Start the motor.
9. Measure the running current at:
   A. L1＿＿＿＿＿A.
   B. T1＿＿＿＿＿A.

   Explain the reason for the readings.
10. Extend time delay to approximately 60 seconds.
11. Start the motor.
12. Measure the voltage across the:
    A. L1 and L2 terminals＿＿＿＿＿V.
    B. T1 and T4 terminals＿＿＿＿＿V (before the preset time-delay contact closes).
    C. T1 and T4 terminals＿＿＿＿＿V (after the preset time-delay contact has closed).

    Explain the voltage readings.
13. Is the time-delay contact released by a TR coil or is it released (triggered) by another contactor?
14. Measure:
    A. Starting current＿＿＿＿＿A.
    B. Changeover current＿＿＿＿＿A.

15. Is this particular wye-delta circuit open or closed transition?
16. List the coil sequence.
17. Is this circuit NVR or NVP?
18. Is wye-delta starting reduced voltage or reduced current starting?
19. When the complete circuit is fully understood, attempt to troubleshoot the circuit.

For troubleshooting information, see introduction to Lab nos. 21 to 26, "Troubleshooting."

CONCLUSIONS

LAB COMPLETED _____ **TEACHER** _____

LAB NO. 23

LAB NO. 24
TWO-STAGE PRIMARY
RESISTANCE STARTING

Physical layout

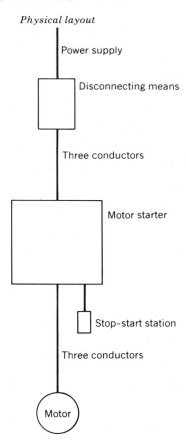

Power supply

Disconnecting means

Three conductors

Motor starter

Stop-start station

Three conductors

Motor

Power circuit—schematic diagram

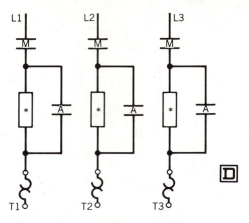

* Resistors

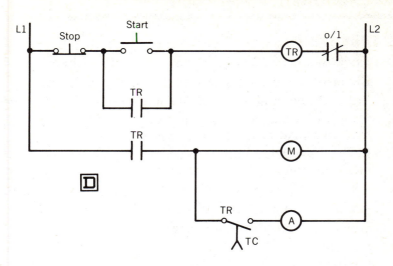

INSTRUCTIONS

1. Lock the disconnecting means in the off position.
2. Record the motor starter ratings and the motor data.
3. Study the general layout of the interior of the motor starter.
4. Using the schematic diagram, locate every terminal, contact, overload device, and coil in the motor starter.
5. Is the time-delay contact released by a TR coil, or is it released (triggered) by another contactor?
6. Select the correct overload heater for this installation.
7. Adjust the time-delay contact for 15 seconds.
8. Unlock the disconnecting means.
9. Start the motor.
10. Measure:
    A. Starting current_____A.
    B. Changeover current_____A.
11. Measure:
    A. Starting voltage_____V.
    B. Running voltage_____V.
    (Measurements for no. 11 to be taken across T1 and T2.)
12. List coil sequence.
13. Is the starting method open or closed transition?
14. Is the installation NVR or NVP?

15. When the complete circuit is fully understood, attempt to troubleshoot the circuit.

For troubleshooting information, see introduction to Lab nos. 21 to 26, "Troubleshooting."

CONCLUSIONS

LAB COMPLETED _____ **TEACHER** _____

LAB NO. 25
AUTOTRANSFORMER STARTING

Physical layout

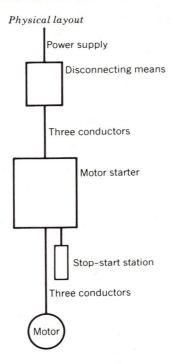

Power supply

Disconnecting means

Three conductors

Motor starter

Stop-start station

Three conductors

Motor

Power circuit—schematic diagram

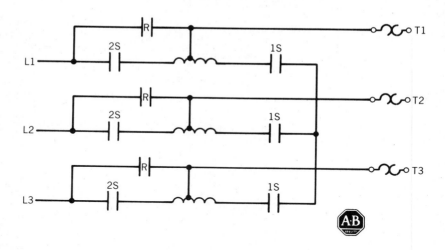

Control circuit—schematic diagram

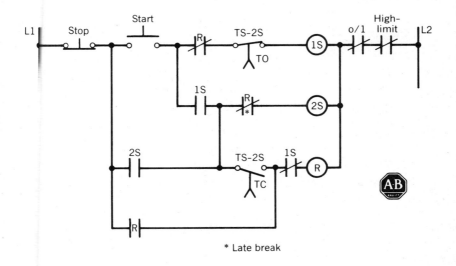

* Late break

INSTRUCTIONS

1. Lock the disconnecting means in the off position.
2. Record the motor starter ratings and the motor data.
3. Study the general layout of the interior of the motor starter.
4. Using the schematic diagram, locate every terminal, contact, overload device, and coil in the motor starter.
5. Select the correct overload heater for this installation.
6. Adjust the time-delay contacts for 60 seconds.
7. Disconnect the motor leads: T1, T2, and T3. Do not reconnect the motor leads until requested by these instructions.

   Note: Each autotransformer will have three percentage taps: two insulated and one terminated on a binding post. *Follow instructions 8 to 17 very closely.*

8. Unlock the disconnecting means and turn it to the on position.
9. Press the start button.
10. Measure the voltage between T1 and T2 (before the time-delay contacts operate). The percentage tap was _____ and _____ V.
11. Press the stop button and turn off the disconnecting means.
12. Change the autotransformer taps and insulate the unused taps.
13. Repeat instructions 8, 9, 10, and 11.
14. Change the autotransformer taps, using the third set, and insulate the unused sets.
15. Repeat instructions 8, 9, 10, and 11.
16. Compare the measured voltages and percentage taps with the comments about autotransformer percentage taps in Section 6.
17. Reconnect the motor leads to T1, T2, and T3.
18. Adjust the time-delay contacts for six seconds.
19. Measure:
    A. Starting current_____A.
    B. Changeover current_____A.
    C. Running current_____A.

20. Are the time-delay contacts released by a TR coil or are they released (triggered) by another contactor?
21. Is this circuit open or closed transition?
22. Is this circuit NVR or NVP?
23. List coil sequence.
24. When this circuit is fully understood, attempt to trouble-shoot the circuit.

Fcr troubleshooting information, see introduction to Lab nos. 21 to 26, "Troubleshooting."

CONCLUSIONS

LAB COMPLETED _____ **TEACHER** _____

LAB NO. 25

LAB NO. 26
TWO-STAGE
SECONDARY RESISTANCE STARTING

Physical layout

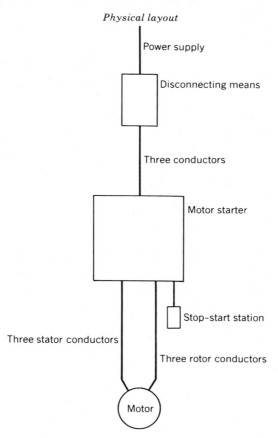

Power supply

Disconnecting means

Three conductors

Motor starter

Stop-start station

Three stator conductors

Three rotor conductors

Motor

Power circuit—schematic diagram

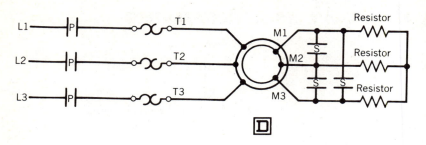

Control circuit—schematic diagram

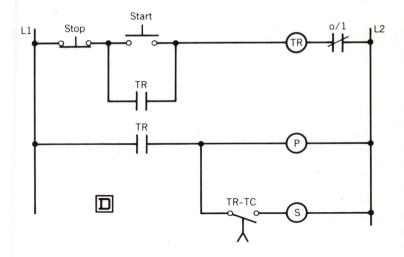

INSTRUCTIONS

1. Lock the disconnecting means in the off position.

2. Record the motor starter ratings and the motor data.

3. Study the general layout of the interior of the motor starter.

4. Using the schematic diagram, locate every terminal, contact, overload device, and coil in the motor starter.

5. Select the correct overload heater for this installation.

6. Adjust the time-delay contact for 60 seconds.

7. Connect a voltmeter across T1 and T2, and another voltmeter across M1 and M2. Set the voltmeter scales at least equal to the line voltage.

8. Unlock the disconnecting means and turn it to the on position.

9. Press the start button and record the voltage between T1 and T2_____V.

10. Record what happened to the measured voltage between M1 and M2:

    A. Prior to the closing of the time-delay contact.

    B. After the time-delay contact closed.

11. Use a tachometer to check motor speed:

    A. With the external resistance in the rotor circuit_____rpm.

    B. With the external resistance by-passed_____rpm.

12. Is this circuit open or closed transition?

13. Is this circuit NVR or NVP?

14. List coil sequence.

15. When the circuit is fully understood, attempt to trouble-shoot the circuit.

For troubleshooting information, see introduction to Lab nos. 21 to 26, ''Troubleshooting.''

CONCLUSIONS

LAB COMPLETED _____ **TEACHER** _____

CONTENTS

LAB SECTION B
DC MOTOR CONTROL LABS

LAB SECTION B

DC MOTOR CONTROL LABS

INTRODUCTION TO DC LABS

Labs 27 to 31 have been designed to allow the student to observe and understand a few basic concepts about the operation and control of DC motors. Many more labs could be prepared to teach the student how to coordinate the physical layout, schematic diagrams, and the wiring diagram. The teacher should expand the suggested workstation to make the DC labs more meaningful and interesting.

MOTOR CONTROL WORKSTATION

The motor control workstation shown in Figure B-1 is one suggested layout. A layout has been provided so that the physical layout illustrated for each lab will relate to an actual installation.

CONTROL COORDINATION

All components of a DC control system must be properly coordinated to suit the motor to be operated. The haphazard selection of DC control equipment could cause equipment failure, coil burnout, and so on.

LAB INSTRUCTIONS

Each lab has been divided into several parts:

Physical layout.
Wiring diagram.
Schematic diagram.
Instructions.
Conclusions.

Physical Layout

The physical layout relates to the motor control workstation, shown in Figure B-1. This layout should assist the student to relate to a motor control installation.

1. 15-A two-pole circuit breaker.
2. Wiring trough, 4 in. × 4 in.
3. Power supply terminal strip.
4. Magnetic across-the-line motor starter.
5. Start–stop remote pushbutton station.
6. No field release relay (field loss relay).
7. CEMF relay.
8. Variable field resistance.
9. Variable series resistance.
10. Control relay.
11. Pneumatic time-delay relay.
12. DC motor.
13. Anti-plugging relay.
14. Reversing magnetic contactor (hp rated).
15. Forward–reverse–stop remote pushbutton station.
16. Wiring trough, 4 in. × 4 in.

Wiring Diagram

The wiring diagram relates to the actual position of each component of the control system.

Schematic Diagram

The schematic diagram will match the wiring diagram, but will be easier to read.

Instructions

The instructions will encourage the student to obtain as much information as possible from each lab.

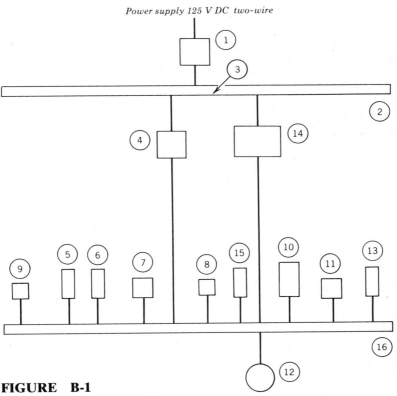

FIGURE   B-1
Motor Control Workstation.

Conclusions

The conclusions will include a summing up of the lab. Electrical code requirements (if applicable) should be recorded.

Evaluation of Labs

The teacher may evaluate each lab at the workstation or request lab reports to be submitted for evaluation.

Answers, conclusions, and completed drawings for all labs will be located in the Instructor's Manual.

GENERAL INSTRUCTIONS FOR LAB NOS. 27 TO 31

1. Indicate the number of conductors required and the colors chosen on the physical layout provided for each lab.
2. Install the conductors. Complete all connections. Operate the equipment.
3. Perform all requirements for each lab.
4. Record any relevant data.
5. Answer all questions.
6. All wiring is to be installed in a neat and professional manner.
7. Do not abuse the equipment.
8. Have the installation and conclusions evaluated by the teacher.

LAB NO. 27
MAGNETIC ACROSS-THE-LINE
MOTOR STARTER TO OPERATE A
SHUNT MOTOR, THREE-WIRE CONTROL

Physical layout

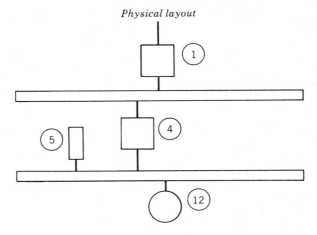

Schematic diagram

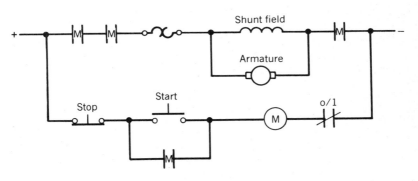

INSTRUCTIONS

1. Connect the circuit as shown by the schematic diagram.
2. Operate the equipment.
3. Compare the lab operation with the lesson on the topic in Section 11.
4. Select the correct overload heater for this installation.
5. Is this circuit NVR or NVP?

CONCLUSIONS

LAB COMPLETED _____ **TEACHER** _____

© 1987 by John Wiley & Sons, Inc.

LAB NO. 28
NO FIELD RELEASE
CIRCUIT ADDED TO LAB NO. 27

Physical layout

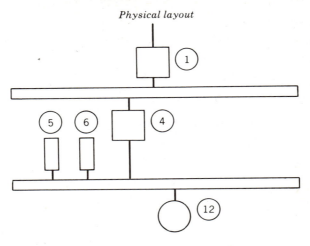

Schematic diagram

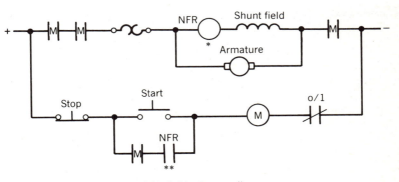

* No field release coil
** No field release contact

INSTRUCTIONS

1. Connect the circuit as shown by the schematic diagram.
2. Operate the equipment.
3. Compare the lab operation with the lessons on the topic in Section 11.
4. Reverse the direction of rotation of the motor.

CONCLUSIONS

LAB COMPLETED _____ **TEACHER** _____

LAB NO. 29
TWO-STAGE REDUCED VOLTAGE STARTING TO OPERATE A COMPOUND MOTOR

Physical layout

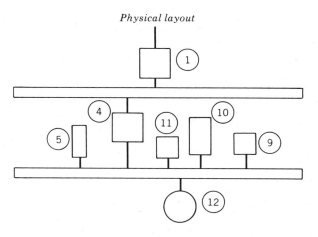

Schematic diagram

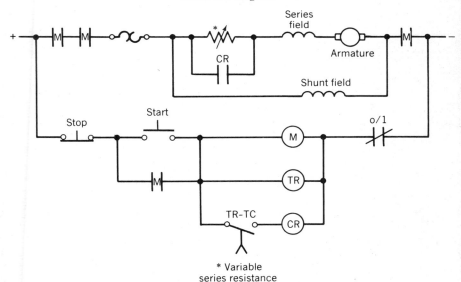

* Variable
series resistance

INSTRUCTIONS

1. Connect the circuit as shown by the schematic diagram.
2. Operate the equipment.
3. Compare the lab operation with the lessons on the topic in Section 11.
4. Adjust the time-delay contact for 60 seconds.
5. Record the below-base speed (maximum resistance)___rpm.
6. List coil sequence.
7. Adjust the time-delay contact for three seconds.
8. Measure:
   A. Starting current_____A.
   B. Transition current_____A.
   C. Running current_____A.
9. Connect a voltmeter across the series resistance and another voltmeter across the armature. Start the motor and observe the relationships through the starting and running period.

CONCLUSIONS

LAB COMPLETED _____   **TEACHER** _____

LAB NO. 30
OPERATING A DC
MOTOR ABOVE BASE SPEED

Physical layout

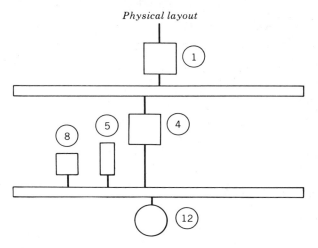

Schematic diagram

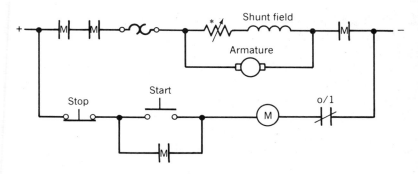

INSTRUCTIONS

1. Connect the circuit as shown by the schematic diagram.
2. Operate the equipment.
3. Compare the lab operation with the lessons on the topic in Section 11.
4. Record:
   A. Base speed_____rpm.
   B. The maximum above-base speed with the installed field rheostat_____rpm.
5. Explain what would happen if the field rheostat should open-circuit while the shunt motor was running.

CONCLUSIONS

LAB COMPLETED _____   TEACHER _____

LAB NO. 31
ANTI-PLUGGING

Physical layout

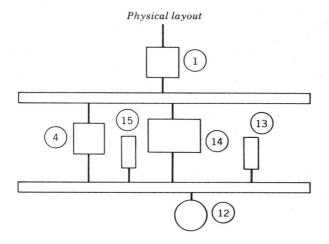

Schematic diagram

INSTRUCTIONS

1. Complete the installation using the schematic diagram shown in Figure 11-43.
2. Apply the comments in Section 11 on anti-plugging to the installed circuit.
3. Use this circuit to practice troubleshooting.

CONCLUSIONS

LAB COMPLETED _____ **TEACHER** _____

© 1987 by John Wiley & Sons, Inc.